MW00826965

Franco and the Condor Legion

Franco and the Condor Legion

The Spanish Civil War in the Air

Michael Alpert

BLOOMSBURY ACADEMIC
LONDON • NEW YORK • OXFORD • NEW DELHI • SYDNEY

BLOOMSBURY ACADEMIC
Bloomsbury Publishing Plc
50 Bedford Square, London, WC1B 3DP, UK
1385 Broadway, New York, NY 10018, USA

BLOOMSBURY, BLOOMSBURY ACADEMIC and the Diana logo are trademarks of Bloomsbury Publishing Plc

First published in Great Britain 2019

Cover images: Francisco Franco (1892–1975), circa 1937 (Photo by Hulton Archive/Getty Images). German Heinkel He-111 aircraft of Condor Legion bombing the town of Valencia, Spain, during the Spanish Civil War, *c.*1938–39.
(Photo by ullstein bild/ullstein bild via Getty Images)
Cover design: Paul Smith Design

A catalogue record for this book is available from the British Library.

A catalog record for this book is available from the Library of Congress.

ISBN: HB: 978-1-7883-1118-2
ePDF: 978-1-7867-3563-8
eBook: 978-1-7867-2563-9

Typeset by Newgen KnowledgeWorks Pvt. Ltd., Chennai, India
Printed and bound in Great Britain

Contents

Illustrations

Preface

On 22 May 1939, several thousand German *Luftwaffe* (air force) personnel paraded on the aerodrome of the northern Spanish city of León in the presence of General Francisco Franco, the victor in the civil war which had come to an end a few weeks earlier on 1 April. The parliamentary and democratic Spanish Republic had been bloodily defeated, to be replaced by a repressive dictatorship which would last until Franco's death in November 1975.

Adolf Hitler's Nazi Germany had aided Franco's cause by providing him with an expeditionary air force called the Condor Legion. At the parade, as the military bands crashed out German and Spanish marches, Franco, accompanied by the most important Insurgent, or Nationalist, as they preferred to be called,[1] commanders, together with the ambassadors of Germany and of Fascist Italy, which had also sent massive aid to Franco, gave warm speeches of appreciation and farewell. The last commander of the Condor Legion, General Wolfram Freiherr (Baron) von Richthofen, a cousin of the famous air ace of the First World War, speaking in Spanish, replied referring to the 'shared shedding of blood, which has deepened and eternalised German – Spanish friendship', as he presented a collection of one million pesetas for the families of Spanish pilots who had lost their lives on Franco's side.[2] From León the 5,136 Germans still in Spain of the 19,000 who had at some time served there, drove the 250 miles to the port of Vigo, where they marched proudly through the city to the cheers of the spectators and drank one last glass of Spanish wine before embarking in ships of the official Nazi leisure and tourism organization, known as *Kraft durch Freude* ('Strength through Joy'). Setting sail for Germany on 25 May, they took with them the most modern of the aircraft that had helped to win victory for Franco: Messerschmitt Bf-109E fighters, Heinkel-111 bombers and Junkers-87 dive-bombers, though most out-of-date equipment was left and would equip the Spanish air force for many years to come.

They arrived to a heroes' welcome in Hamburg and were taken to the major base of Döberitz, where they had first been mustered in July 1936, before reassembling for a final triumphant parade in Berlin on 6 June 1939 in which fourteen thousand men of the Condor Legion took part. Then they returned to the Luftwaffe and the other units from which they had been selected to help

Franco save Spain from the 'Bolshevism' which propaganda had alleged was about to overwhelm traditional and Catholic Spain, whose older values were now to be restored by Franco and his heroic 'Glorious National Movement'.

The Spanish Civil War was fought on land and at sea but also in an age of great interest in air warfare and the very rapid development of warplanes. In the latter 1930s, one-gun, wooden and fabric biplanes, not too different from those of the First World War, gave way to the monoplane, all-metal, heavily armed British Spitfire and Hurricane, and the German Messerschmitt Bf-109, which would fight the battle of Britain in 1940. Biplanes might be very manoeuvrable, but the monoplane had the advantage of superior speed and wings strong enough to hold more and heavier guns than the biplane as well as to withstand the much higher speeds provided by more powerful engines. The Spanish war began with the biplanes of the late 1920s with which the Spanish *Aviación de Guerra* or air force was equipped and ended after two years and eight months with state-of-the-art machines supplied and flown by German and Italian pilots in the service of Franco's Nationalists, as well as Soviet machines and Russian pilots who had been sent to Spain to save the Republic from impending defeat. Germany, Italy and the Soviet Union or USSR sent the two warring Spanish sides Heinkel, Junker, Dornier, Messerschmitt, Savoia-Marchetti, Fiat, Polikarpov and Tupolev warplanes. These were the most advanced bombers and fighters of their time, whose makes would become familiar during the Second World War. In Spain, they were to be tried out in what is often referred to as a rehearsal for a major European war. The participating German, Italian and Soviet air forces studied the experiences of pilots and the performance of the various machines, while they tried to draw lessons from Spain which they could apply in a future conflict. The vital questions which the Great Powers had to face in the latter 1930s were the following: how far was the British Prime Minister, Stanley Baldwin, right when he declared to the House of Commons in 1932 that 'the bomber will always get through'?[3] Should countries rely on large bomber forces to destroy their enemies, as in the 1936 film *Things to Come*, or should better-armed, more manoeuvrable and faster fighter aircraft be developed? Could the Spanish Civil War provide reliable answers?

Europe in 1936

By 1936 when the Spanish Civil War broke out, Britain and France, the European victors of the 1914–18 war, were having to deal with problems created by the accession to power in Germany of Adolf Hitler and the Nazis.[4] Hitler had become Chancellor in January 1933. He had rejected the terms of the Versailles Treaty of 1919 which had ended the First World War. He had recreated the banned German general staff and the air force, now called the *Luftwaffe*. In June 1935, by the Anglo-German Naval Treaty, Britain had recognized the right of Germany to create a sizeable navy. In March 1936, France and Britain had silently consented to German troops marching into the demilitarized zone of the Rhineland.

As for Fascist Italy, in 1935–6, Britain and France had had to accept the conquest and establishment by the dictator Mussolini of a colonial regime in Abyssinia. As the Spanish Civil War was beginning, in July 1936, the last sanctions which had been imposed on Italy for its aggression by the League of Nations were being lifted. Italy and Germany were spending vast amounts of money on armaments while British and French diplomatic efforts were directed to avoiding, at almost all costs, another major European war.

The USSR, for its part, was just emerging from being treated as an international pariah after the Russian revolution had confiscated all private property and murdered the Russian royal family. The West deeply and widely feared communist subversion, while Hitler's rhetoric was consistently anti-Soviet and threatening. Consequently, in August 1935, Moscow had announced a new policy. It would back-pedal its aim of world revolution and would henceforth instruct communist parties to cooperate in 'Popular Fronts', that is, electoral alliances with all forces, even conservatives, provided that they recognized the danger posed by Nazis and Fascists, whose powerful movements were threatening to overthrow liberal and parliamentary regimes. In France, where such extremist movements were menacingly strong, the Popular Front alliance had won the elections of June 1936. When the Spanish Civil War broke out in July, France was governed by the socialist Léon Blum, but in the previous year Britain had elected an overwhelming Conservative Party majority under Stanley Baldwin. Blum's position on Spain would be weakened by the French insistence on not diverging from British policy. While Blum would want to support the Spanish Republic, Baldwin would strive to keep Britain completely neutral.

As for Spain, it was a backward country, which had been neutral in the First World War. It counted for little in European politics. Spanish experience of war had been limited to its slow suppression or 'pacification' of rebellious tribes in the zone of the Moroccan Protectorate which had been allocated to Spain by the 1912 Treaty of Fez. While the Spanish air force had played an important role in bringing supplies to remote desert posts and had carried out massive air bombing against villages, souks, water supplies and cattle, especially with mustard gas, and had developed techniques of firing, aiming and bombing over at least six years,[5] it had never had to fight an enemy in the air and had not developed modern air war techniques.

Political, economic and social advance in Spain had been slow. However, the country had undergone a peaceful revolution when the king, Alfonso XIII, abdicated on 14 April 1931 in favour of a Republic with progressive ideals. Since then, extensive reforms had been implemented, though not very successfully, in landownership, the separation of Church and State and the role and structure of the army. However, conservative and traditional hostility to progressive change had been great, and this had led to two years, between November 1933 and February 1936, of reaction and standstill in the reforms. In February 1936, a modified Popular Front had won a narrow electoral victory over a conservative opinion alarmed at the apparent strength of the three great Spanish working-class organizations, the anarcho-syndicalist *CNT* or *Confederación Nacional de Trabajo* (National Labour Federation), the socialist *UGT* or *Unión General de Trabajadores* (General Workers' Union) and its political party, the *PSOE* or *Partido Socialista Obrero Español* (Spanish Socialist Workers' Party), all of which seemed, to conservative opinion, to be planning revolution with the aid of progressive political parties.

In Spain the army, in which air force officers had begun their careers, had traditionally engineered coups to overthrow civilian governments. The Republic of 1931 had been preceded by the dictatorship of General Miguel Primo de Rivera. In 1932 army officers had attempted a coup, while since the victory of the Popular Front in February 1936 officers in garrisons all over Spain had been planning to declare martial law, overthrow the government and replace it either with an authoritarian Republican regime or bring back the monarchy. A few officers were members or sympathizers of the small but violent Fascist *Falange* party.

The army uprising which led to the Spanish Civil War began on the afternoon of Friday, 17 July 1936, in Tetuán, capital of the Spanish zone of the Protectorate of Morocco, and was successful there within thirty-six hours, crushing all resistance. On Saturday, 18 July, Franco, who was the general in command in the Canary Islands, proclaimed martial law as a preliminary to the 'Glorious National Revolution' and called by radio on all Spanish garrisons to come out in rebellion. Many did so. In Barcelona, however, the insurgents were defeated on Sunday, 19 July. They were crushed in Madrid on Monday, 20 July, as well as in many other towns and cities in central, southern and eastern Spain, in Catalonia, in the Basque Country and a fringe along Spain's northern coast from the French frontier through the Basque Country and the provinces of Santander and Asturias, as far as Oviedo. However, the insurrectionary garrisons were successful in large parts of western and northern Spain and in large and smaller cities, including Zaragoza, Seville, Córdoba and Granada, and they were able to take southern ports such as Cádiz and Algeciras. Thus the success of the uprising in some parts of the country, in contrast with the failure to rebel or of rebellions of some garrisons in other parts, often led by officers loyal to the Republic, plus the fierce resistance in the larger cities of political parties of the left and centre, together with workers' groups, meant that Spain was to be faced by a civil war.[6]

The only general works available on the Spanish Civil War in the air are General Jesús Salas Larrazábal's *La guerra de España desde el aire* (Esplugues de Llobregat, 1969) and his enormous *Guerra Aérea 36/9* (Madrid, 1998). The former deals almost entirely with fighter operations and, though it has appeared in English,[7] it is often unmanageable because of its breathless accumulation of detail, which is also characteristic of Salas's later book. The latter is nevertheless valuable because of its extreme detail. In English, Gerald Howson's *Aircraft of the Spanish Civil War* (Putnam, 1990) is basically a thorough list of aircraft and their operations but not organized chronologically. The present book, while laying emphasis on the German contribution to Franco's victory, describes and analyses in chronological order and depth the intervention of German, Italian and Soviet aircraft in the Spanish conflict, as well as the supply of aircraft in general and the role of a number of volunteer and mercenary airmen, while trying to explain the result of the war, its importance for the Second World War and the possible lessons learnt.

Map of Spain with places mentioned in the text

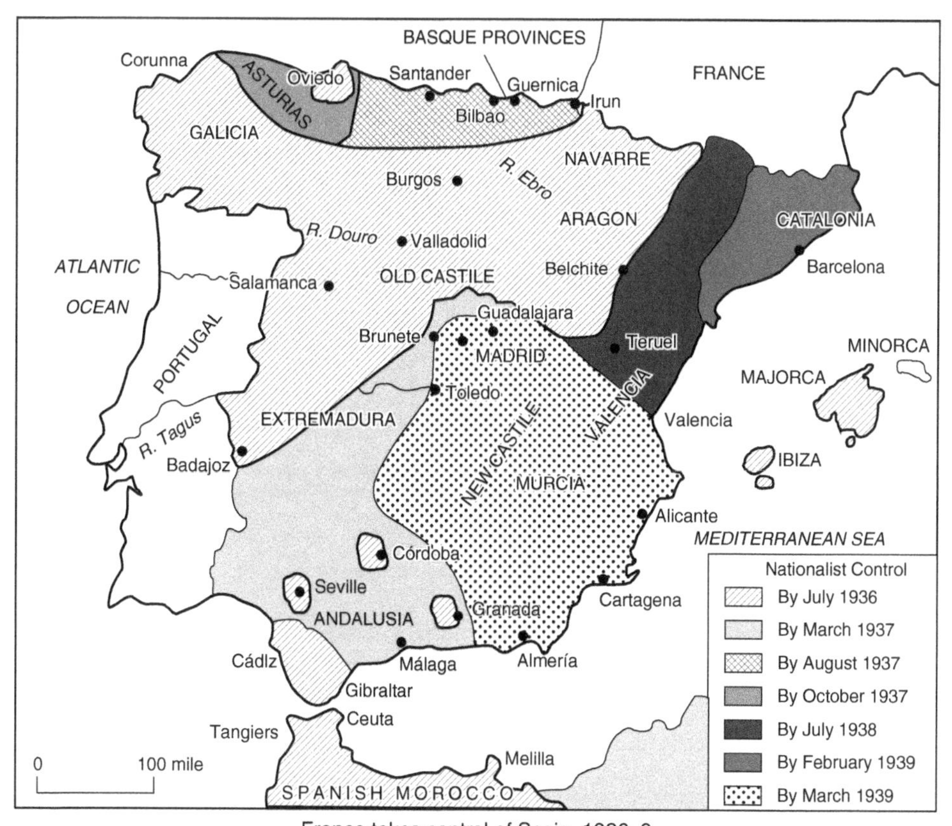

Franco takes control of Spain, 1936–9

Chapter 1

The Spanish Civil War begins: A disaster which brought Franco to power

Some individual flights, fortunate or disastrous, which took place at the very outset of the Spanish Civil War, made a substantial difference to the outcome. They facilitated the meteoric rise of General Francisco Franco, who had been a lesser figure in the planning of the insurrection which led to the war, and led to his eventual dictatorship which lasted till his death in November 1975, over forty years after he declared military rebellion on 18 July 1936.

General José Sanjurjo, known as the 'Lion of the Riff' for his military triumphs during the pacification of the Spanish zone of Morocco in the 1920s, was to be the leader of the uprising planned for July 1936. In August 1932 he had led a coup against the reforming Spanish Republic, which had been established on 14 April 1931 after King Alfonso XIII had abandoned the throne. Sanjurjo was a leading figure in the army, both among those who sought a restoration of Alfonso XIII and for the Carlists or Traditionalists. The latter supported the claims of a pretender in a dynastic squabble going back to the nineteenth century. In 1936, they still constituted a major force which would be of great military aid to the uprising of July 1936 and the subsequent civil war.

Unsupported by most of his fellow officers, Sanjurjo's 1932 coup failed. Fleeing towards Portugal, the general surrendered on the frontier. He stood trial and was condemned to death for military rebellion, but President Manuel Azaña, perhaps mindful of the executions of army officers who had mounted an abortive Republican coup in 1930 and had achieved the status of martyrs, commuted Sanjurjo's sentence to life imprisonment. However, having been dismissed from the army, the distinguished general would have to serve his sentence in a convict gaol rather than in the less shameful conditions of a military prison reserved for officers. He had to wear prison overalls and live among robbers and murderers, which surrounded him with a halo of martyrdom in the eyes of many. The victory of right-wing parties in the elections of November 1933 was followed by amnesty for Sanjurjo the following April, after which the general went into exile in the conservative and undemocratic *Estado Novo* of Portugal presided over by António d'Oliveira Salazar.

At Spanish elections on 18 February 1936, a victorious alliance of centre and left parties, in a weak imitation of the Popular Front which would triumph in France in May and June 1936, induced several groups, among them officers, the young Fascist-like Falange movement and a spectrum of conservatives and monarchists, to begin serious planning for the insurrection which many of them had been considering ever since the Republic had been established in 1931. The murder in Madrid, during the torrid night of 12–13 July 1936, of the leader of the right-wing National Front, José Calvo Sotelo, carried out by left-wing members of the armed police, in revenge for the assassination of left-wing officers who had been advising working-class groups how to defend themselves against the feared army uprising, acted as the touchpaper to the declaration of military law by army garrisons over most of Spain.

The plan was for General Sanjurjo to be flown from his exile in Portugal to the headquarters of the Burgos army division, in the right-wing area of northern Castile. From there, as a unifying figurehead, he would lead the military uprising which aimed to march on Madrid, overthrow the Popular Front government of the Republic and replace it with an authoritarian right-wing regime.

From his command at Pamplona, capital of Navarre, General Emilio Mola, the planner of the details of the uprising, sent a De Havilland Puss Moth biplane, one of three single-passenger light sports and touring aircraft recently acquired by Spanish purchasers, to Lisbon on Sunday, 19 July 1936. The ardent royalist and experienced aviator Major Juan Antonio Ansaldo was to fly Sanjurjo from his Portuguese exile to Burgos where, since the little Puss Moth had no radio, white sheets, spread out on the ground of the aerodrome, would indicate that the garrison had successfully imposed martial law.

The Portuguese government, wary of complaints from neighbouring Spain that Ansaldo was planning to take off with the distrusted and exiled general from a military aerodrome, instructed him to pick Sanjurjo up from a smaller landing strip, in fact a racecourse, at Marinha, on the south coast of Portugal, so that Lisbon could state officially that it was unaware of the general's departure for Spain. This was to have disastrous consequences for Sanjurjo, but it would be a step in bringing Franco nearer to power.

On the following day, Monday, 20 July 1936, the direction of the wind at Marinha obliged Ansaldo to take off in the direction of a grove of pine trees. General Sanjurjo insisted on loading the Puss Moth with a trunk containing the ceremonial uniform which he looked forward to wearing when he was announced as the Saviour of Spain. 'I need to wear proper clothes as the new *caudillo* of Spain', he claimed, or so Ansaldo recounted in his memoirs.[1] Later, Franco would use this medieval Spanish title, corresponding to the Italian *Duce*, the German *Fuehrer* and, far away in England, the absurd-sounding 'Leader', a title by which the head of the British Fascist movement, Oswald Mosley, was known. It was precisely the weight of Sanjurjo's baggage which was thought at the time to have been the cause of the disaster which killed him, though it did not prevent the Puss Moth from taking off.

Ansaldo, nervous but at the same time recalling how many risky take-offs he had carried out successfully, taxied to the far edge of the field, turned, revved his motors to their maximum and surged forward over what he recalled later as bumpy ground. At a speed of 10 mph faster than normal, to ensure he could clear the trees, he felt a heavy knock. He took off, but the aircraft was vibrating intensely. Ansaldo thought he had damaged his propeller. Urgently seeking a place to land before the propeller shattered, he was unable to clear a stone wall. The fully fuelled aircraft burst into flames. Ansaldo, his flying kit on fire, managed to open the door and throw himself out, while the general, his skull fractured and perhaps already dead, was swallowed up in the conflagration. Later, there would be unfounded rumours that the ambitious Franco had arranged for the Puss Moth to be sabotaged. This is unlikely, because Franco, though admittedly wily, was quite junior among the leaders of the uprising and not even in the running to become, as he later would, generalissimo, head of the Nationalist government and Head of State.

A successful flight for Franco

Generals of division, equivalent to British major generals, held the highest rank in the Spanish army, but Francisco Franco was almost the most junior of them. He was military commander of the Canary Islands where he was thought by the Spanish government to be far enough away from the mainland not to be involved in military coups, despite the suspicion in which his monarchist background was held. His planned role in the uprising was to lead a military insurrection in the Spanish zone of northern Morocco, where he had served with distinction during the Riff wars of the 1920s and reached the rank of brigadier general at the very early age of 34. He was immensely influential in the officer corps, having been director of the General Military Academy and chief of staff during the military suppression of a violent left-wing uprising in the mining districts of Asturias in October 1934.[2] Thus it was vital to bring him from his headquarters in the Canaries to Tetuán, capital of Spanish Morocco, where he would take command of the Moorish professional troops, known as the *Regulares*, and the Legion, a disciplined, well-trained and brutal colonial-type force created on the model of the French Foreign Legion but mostly Spanish in composition.

To bring Franco undetected from the Canary Islands to Morocco demanded speed and concealment. The journey required an aircraft which could cover the 850-mile distance quickly and with the minimum of refuelling stops. No such private machine was available to the conspirators in Spain, so arrangements were made to charter a multi-seat state-of-the art passenger aircraft. This was a De Havilland Dragon Rapide DH-89A belonging to Olley Air Service at Croydon Airport, south of London. This aircraft, one of the most successful piston-engined machines ever built, could accommodate six to eight passengers depending on the fuel load. It had a range of 578 miles and could thus reach Tetuán from the Canaries with only one or two stops.

The plane was chartered by Luis Bolín, London correspondent of the Spanish monarchist newspaper *ABC*. The money for the charter was provided by the multimillionaire Juan March through a London bank. Leaving Croydon at 7.15 am on Saturday, 11 July 1936, the Dragon Rapide was piloted by Captain Cecil Bebb, who later claimed (in a 1983 Granada Television documentary) that he had thought that his plane was going to liberate a Riff chieftain from exile in the Canaries and fly him back to Morocco. This is an extraordinary indication of the attitude of pilots such as Captain Olley, director of Olley Air Service, and Bebb, who did not seem to be aware that they might have caused a serious international incident through interfering in the internal affairs of another country. The flight simulated a tourist jaunt to the Canaries and carried, beside the pilot, radio operator and flight engineer, two attractive young women as cover in case the Spanish government's spies were watching their fellow passenger Bolín, who had published a book in Britain three years earlier attacking the Spanish Republic.[3] The most intriguing of the travellers, however, was the father of one of the young women. This was Major Hugh Pollard, who had been in War Office Intelligence and had experience of the Troubles in Ireland and revolutions in Mexico. He was suggested by Douglas Jerrold, who had served in British Intelligence and was now a right-wing publisher and a fellow author of the book written by Bolín attacking the Spanish Republic.[4] Pollard's presence on the flight suggests that MI6 or some similar organization may have had knowledge of the mission to bring Franco to Morocco.

The flight was carefully prepared. After stops at Bordeaux, Biarritz, Oporto, Lisbon, Casablanca and in the Spanish Sahara, the Dragon Rapide reached Gando aerodrome on Grand Canary on 14 July. Franco made arrangements to leave his headquarters on Teneriffe and travel to Grand Canary to attend the funeral of the local commander, General Balmes, who had accidentally shot himself while cleaning his revolver. Some historians suspect that Balmes was conveniently killed to give Franco an excuse to apply to Madrid for permission to leave his post and go to Grand Canary on the overnight boat for the funeral, arriving early on Friday, 17 July.[5]

The rising in Morocco was scheduled to begin the following day. At 5 am on Saturday, 18 July, Franco signed the declaration of martial law and set off for Gando airfield, whence Bebb took off with his passenger at 2 pm to fly to Morocco. After landing at Agadir to refuel, they flew on to Casablanca, where they spent the night. Bebb took off next morning for Tetuán. Franco instructed Bebb to circle low over Sania Ramel military aerodrome; he recognized officers whom he knew, thus being reassured that the uprising had been a success. Having deposited his passenger, Bebb flew the Dragon Rapide on to Lisbon where Bolín conferred with General Sanjurjo before the latter's fatal air crash on Monday, 20 July. From there they went to Marseilles. The Dragon Rapide returned to Croydon while Bolín continued to Rome where he had been instructed to ask the Italian government for support for the Spanish military insurrection.

This British aircraft, registered G-ACVR, had played an irreplaceable role in assuring Franco's rise to power. It continued flying until 1953. In the 1960s it was restored, presented to the Spanish government and placed on display at the Museo del Aire.[6] Cecil Bebb, who had begun his flying career in 1921 as a 16-year-old Royal Air Force apprentice and had become a sergeant pilot, went on to become operations manager for British United Airways and to enjoy a distinguished career in British aviation.

Two Spanish generals fly to their deaths before the firing squad

General Manuel Goded, Inspector General of the Spanish army and highly respected for his intellectual qualities, was linked with the reactionary army officers' league, the *Unión Militar Española*. His task in the planned insurrection was to fly from Majorca, where he was commanding officer, to Barcelona, arriving at about midday on Sunday, 19 July, in order to take over the leadership of the uprising among the numerous regiments in barracks in the city. Summoned by Goded, five Spanish air force Savoia-Marchetti S-62 seaplanes arrived in Majorca that morning from their base on Minorca. One of the pilots, realizing that he was taking part in an insurrection, flew back to Minorca, which would remain in Republican hands for the rest of the war. At about midday, the remaining four machines with Goded, two officers and Goded's son aboard, swept in over the seaplane base of the great city of Barcelona. They were alarmed not to see the white cross which was to indicate that the insurrection had succeeded. The Catalan regional government, the *Generalitat*, was still in charge despite the army regiments having come out into the streets. Goded had a hard task before him. He signed to his pilot to go down, but even as he was being driven from the seaplane base to military headquarters he was fired on. As the hot July afternoon wore on, it became clear that none of the objectives of the insurgents had been reached. The military uprising had failed because of the concerted action of the organized Barcelona working class and their immense anarcho-syndicalist union, the *Confederación Nacional de Trabajo* or CNT, together with the loyalty of the militarized police force, the *Guardia Civil*, and the air base at El Prat, today Barcelona's civil airport. As night began to fall, General Goded broadcast his surrender to his supporters: 'Luck has been against me [. . .] if you want to avoid more bloodletting, I free you from your oath of loyalty.'[7] He would be tried within a few days by court martial and executed for military rebellion.

General Miguel Núñez de Prado, commander of the air force, also lost his life before a firing squad for the crime of military rebellion, but in his case it was at the hands of the insurgents. The military uprising of 18 July 1936 consisted of the declaration by garrisons all over Spain of the state of war, which put the area concerned under military law. This was against the constitution because only the civil authorities could institute the state of war. However, the army saw itself as bound by the articles of the *Ley Constitutiva del Ejército* or Constitutional Army Act of 29 November 1878, whose Article 2 required the army to defend Spain not only from exterior enemies but from internal ones also. The justification for the unilateral declaration of the state of war, which gave all power to the military, was that the government was no longer able or willing to fulfil its obligations to maintain law and order. This was the official view throughout the Franco years.[8]

Núñez de Prado had been highly decorated and swiftly promoted for valour in the wars of the 1920s in the Spanish zone of Morocco. Already an army colonel, he graduated as pilot and observer in 1923–4. Now a general of division, the highest possible rank under the Second Republic, Núñez de Prado was appointed head of the air force and civil aviation (*Director General de Aeronáutica*) on 12 January 1936. As such, he tried to ensure that officers loyal to the Republic were placed in sensitive posts. When the military insurrection began on the evening of Friday, 17 July 1936, Núñez de Prado assured himself first of the loyalty of the officers commanding the airfields of Barajas, Cuatro Vientos and Getafe, all close to Madrid. Then he was ordered to fly to Morocco and quash the rebellion there, but news came that the airfield at Tetuán was already taken by the insurgents. Another mission was, therefore, selected for him. He was to fly north-east to Zaragoza to persuade the commanding general there, Miguel Cabanellas, who was a known Republican and a Freemason, which in Spain indicated a person of progressive sympathies, not to join the insurrection. At Getafe, to the south of Madrid, accompanied by his military secretary, his aide-de-camp, the pilot and a mechanic, he boarded one of the De Havilland DH-89s of the Spanish LAPE (*Líneas Aéreas Postales españolas*) postal service. Landing at Zaragoza at about 3 pm, he was driven to military headquarters in the city to interview Cabanellas, an ex-comrade in arms in Morocco. Núñez de Prado had bravely ventured into the wolf's lair. He soon realized that the senior officers were hostile and that Cabanellas, with his white beard which made him look years older than his true age, was merely a figurehead. The fate of General Miguel Núñez de Prado has always been a mystery, but he was transferred to prison in Pamplona, possibly 'released' into the hands of extremists and probably assassinated.[9]

Franco's last potential rival dies in an air crash

Brigadier General Emilio Mola Vidal, stationed at Pamplona and 'director' and brains of the insurrection of 18 July 1936, was the only remaining possible rival of Franco, who became generalissimo and Head of State at the end of September 1936. In the following spring, Mola was successfully commanding the assault on the Basque Country and its capital, Bilbao, after several attacks by Franco on Madrid had failed. On 3 June 1937, Mola was flying from Vitoria to Burgos in one of three British-built Airspeed Envoy low-winged monoplane transport aircraft, acquired before the war. A defector had flown it in 1936 from Madrid to Pamplona.[10] The plane in which Mola was flying crashed into a hillside. Rumours that it had been sabotaged in Franco's interest circulated for years.

Thus, General Sanjurjo, the leader of the uprising, and two potential rivals to Franco, Generals Goded and Mola, were killed during or immediately after flights, while Franco himself was flown successfully to take command of the one part of the Spanish army which was efficient, well-trained and ruthlessly led.

The Spanish air force at the start of the civil war

In the Spanish armed forces, the navy was traditionally the most conservative. In July 1936, uprisings by the navy officers had been planned for the two major Spanish naval bases. The insurgent officers had been successful at El Ferrol on the north coast but had failed in Cartagena in the extreme south-east of Spain. Naval officers had agreed not to hinder the transfer of troops by the Francoists from Morocco to the Peninsula. Since that was precisely what they were ordered to do by the Navy Minister and later briefly Prime Minister, José Giral, who had ordered the fleet to leave El Ferrol and Cartagena and blockade the coast of Morocco, the ships' officers were in fact in a state of mutiny.[11]

Many captains, majors and lieutenant colonels in the various garrisons, which were widely spread all over the Peninsula as well as the Balearic and Canary Islands and in the Moroccan protectorate, arrested their own superiors and made the difference between their units remaining loyal to the Republic and declaring against it. At the same time, in the air force, events in the various air bases and airfields also depended largely on the pilots and other personnel.

While military aviation had begun very early in Spain, having been employed in 1913 to drop bombs by hand on unruly Moroccan tribesmen, the Spanish air force had never fought in a European war. The civil war of 1936–9 was thus the first conflict in which Spanish airmen duelled with aircraft flown by other Spaniards, as well as by Germans, Italians and Russians. Nevertheless, Spanish flyers had an admired place among the pioneers of long-distance flights. The 'Plus Ultra', a Dornier Wal flying boat, carried out the first of these long Spanish flights. Leaving Palos de la Frontera on Spain's southern coast on 22 January 1926, its crew, led by Major Ramón Franco, younger brother of the future general and Spanish dictator, touched down at Buenos Aires on 10 February, having flown 6,420 miles in just under sixty hours at an average speed of 106 mph. Soon after, Eduardo González Gallarza, who would end his career as Air Minister in Franco's government, led three Bréguet XIX biplanes in a flight from Madrid to Manila, capital of the Philippines, beginning on 5 April 1926. He and two others covered the distance in thirty-nine days. Later flights pioneered the route from Spain to Fernando Poo, a Spanish possession in the Gulf of Guinea, while two Spanish pilots flew non-stop to Brazil.

The officers of the Spanish air force were largely indifferent to politics, but nevertheless there was somewhat of a progressive tradition among a minority of airmen, who saw themselves as very modern. This became evident at the end of the dictatorship of General Miguel Primo de Rivera in January 1929 and during the indecisive two years which ended with the declaration of the Second Republic on 14 April 1931. Most air force officers were no more than vaguely liberal; a few were clearly right wing, but a handful, stationed mostly at Cuatro Vientos base, south of Madrid, cherished advanced political ideas but had little ability to influence most of their comrades and, despite going on to occupy high office briefly in the Republican government of 1931, failed to affect politics in general.[12]

Among military officers who schemed to overthrow the government of General Berenguer, which followed the downfall of Primo de Rivera, were a number of senior air force officers, most of whom would serve in the Republican air force during the civil war, among them General Miguel Núñez de Prado, who was executed by the insurgents, Lieutenant Colonel Luis Riaño and Majors Ramón Franco (who, unlike the others, would join his brother in rising against the Republic), Angel Pastor, Ignacio Hidalgo de Cisneros, Antonio Camacho and Felipe Díaz Sandino. Their insurrection on 15 December 1930 consisted of taking off from Cuatro Vientos airfield to throw leaflets over the capital. Ramón Franco desisted from his plan to drop bombs on the Royal Palace because the square in front of it was crowded with civilians and children at play. The revolutionary officers returned to Cuatro Vientos and boarded another aircraft in which they fled to seek asylum in Portugal, whence they went to Brussels and later to Paris to join the survivors of the badly planned and uncoordinated Republican insurrection. Later, they waited in the French capital with other anti-monarchist officers and politicians for the Spanish regime to fall and, as they hoped, be followed by the abdication of Alfonso XIII and the declaration of a Republic. When the Republic was declared on 14 April 1931, the revolutionary officers returned to be welcomed by enthusiastic crowds at Madrid's North Station.

Although the air force had expanded significantly during the Riff wars of the 1920s, Spain's parlous economic state had not permitted much advance since then. As Manuel Azaña, Minister for War and Prime Minister of the Spanish Republic between 1931 and 1933 stated on 10 March 1932 during the debate on the budget:

> Without military aviation we are defenceless, because the other nations with whom Spain might go to war have powerful air forces [. . .] Spain is little better than in the cradle in comparison.[13]

Nevertheless, in an epoch when few countries had established an air force independent of the army, Azaña created the Air Corps or *Cuerpo de Aviación*. On 5 April 1933, he established, directly under the Prime Minister's office, the overarching *Dirección General de Aeronáutica*, which included military, naval and civil aviation, probably with the aim of centralizing the air question and removing it from the dead or hostile hands of other ministries, for the Spanish military, like officers in many other armies, tended to be antagonistic to the idea of an independent air force, and to see aircraft as no more than adjuncts to the role of the land army.

However, such changes were visionary and as yet without practical force. For all the modern attitude towards military flying that the Republic seemed to have brought with it, Spain was a poor country. 1931–3, the first two years of the Republic, saw no plan to acquire modern aircraft, and there were many other aspects of Spain's defence which were seen as needing more urgent attention. Although a major programme of rearmament, including orders for 249 new aircraft, was initiated in 1934–5, by the outbreak of the civil war on 18 July 1936, hardly any new material had arrived in Spain. Thus, when the garrisons declared the state of war in 1936, the Spanish air force was, to say the least, poorly equipped. In the first few weeks of the war, the air force on both sides, with a total of 203 functioning warplanes, was to be stretched beyond all its powers in a very large number of missions under the most adverse conditions.

Aircraft

In 1936, the standard bomber and reconnaissance aircraft in the Spanish air force was the two-seater Bréguet XIX sesquiplane (the upper wing was half as long again as the lower), a French model of 1921 built under licence in Spain, though by 1934 the French air force had already withdrawn it from service. It could carry up to 400 kg of bombs, 6 of 11 kg beneath each of its lower wings, and 48 light missiles launched through a bomb chute on the floor in the middle of the fuselage. For defence, the observer/bomb aimer had two 7.7mm Vickers machine guns mounted on a ring. Two hundred Bréguets were built in Spain between 1926 and 1933. They equipped five bomber wings[14] of the air force. There were two squadrons of nine aircraft each at Getafe, to the south of Madrid; three squadrons at León; and a further three at Logroño, covering northern Spain. There was another wing of two squadrons at Tablada, near Seville, the largest airfield in southern Spain, and a further wing at El Atalayón, the airfield at Melilla on Morocco's Mediterranean coast, and there were flights composed of three machines each stationed at Larache and Cape Yubi, on the Atlantic coast of Spanish Sahara. Bréguets were also used for liaison services and for gunnery and bombing schools and could be used to make up the numbers in squadrons when machines were under repair. Perhaps in total there were about 120 Bréguet XIXs in service in Spain, of which about half were captured by the insurgents. The Spanish version was not equipped with front-firing guns, so the observer gunner could bring his machine gun to bear only if his pilot flew in front or sideways on to the attacking machine. Its maximum speed on the level was 100 mph.[15]

Spain's standard fighter plane was the Hispano-Nieuport-52, a French design built under licence in Spain between 1929 and 1931. It had two front-firing 7.92 mm Vickers machine guns synchronized to fire between the propeller blades. Spanish pilots found it heavy on the controls and liable to ground loop on landing. Its maximum speed was about 140 mph. About fifty-six were in service when the civil war began in July 1936 and a number of these were under repair, including seven which the insurgents captured.

In June 1936, three Hawker Fury fighters, bought from Britain, arriving in crates, were assembled and tested. Spain's aviation authority, the *Dirección General de Aeronáutica*, had contracted to build fifty Fury fighters. They had a stated maximum speed of 187 mph. The international agreement not to intervene in the civil war reached by European countries in August 1936 prevented the contract being carried out. Had it been completed, Spain would have begun the conflict with what was at the time considered one of the best fighters in the world, which entered the Royal Air Force and served until it was superseded by the Hawker Hurricane and the Supermarine Spitfire in time for the battle of Britain.

Four Fokker F.VIIb bomber versions of a widely flown commercial aircraft of the early 1930s had been built in Spain and delivered to the Spanish air base at Cape Yubi in the Western Sahara. Five passenger versions of this machine were used by the Spanish commercial airline Líneas Aéreas Postales Españolas, referred to as LAPE, which also possessed four Douglas DC2 airliners.

As for naval aircraft, these were based at San Javier on the coast in the Mar Menor, a salty lagoon in the extreme south-east of Spain, between Murcia and the great naval base of Cartagena. Seven miles to the south was the major air base of Los Alcázares. At San Javier there were three CASA-Vickers Vildebeest squadrons, which were however as yet not equipped with their torpedoes. They could carry a very heavy load of three quarters of a ton of bombs but were extremely slow. There were also twenty-six Dornier-Wal flying boats, but only eight or nine were airworthy. This was a design of 1922, built in Spain under licence in 1929. With a range of 625 miles, they could play a successful role as long-distance reconnaissance and ship-bombing aircraft. There were thirty-five Savoia-Marchetti S-62 flying boats of a 1929 model, built in Spain between 1931 and 1935. At San Javier, there were two squadrons of nine machines each; at Barcelona seven machines, of which four were under repair. At Mahón, capital of Minorca, there were five of these flying boats, together with another five at Marín on the north-west Atlantic coast.

Very few of these aircraft survived the civil war. They were seen almost at once as out-of-date, which is why both sides, the Republic and Franco's insurgent Nationalists, as they insisted on being known, appealed at once for aircraft to other countries.

The air force remains loyal in Madrid, Barcelona, Los Alcázares and San Javier

Major Ignacio Hidalgo de Cisneros y López de Montenegro, an aristocratic air force officer with a distinguished career in Morocco, who was soon to be the commander of the Republic's air force, returned to Spain in autumn 1935 after resigning from his post as an attaché in the embassy in Rome. He writes, significantly:

> The atmosphere in the air force had changed greatly. The friendly comradeship that had kept us so united and had always been the main aspect of the air force had almost completely disappeared. Political tension had infiltrated our ranks and thus most airmen were in one or other political camp.[16]

Appointed on his return from Rome to a non-flying post by the right-wing government in the autumn of 1935, Hidalgo de Cisneros resumed his contacts with officers of progressive sympathies. He was disturbed to see officers of conservative views ostentatiously displaying right-wing newspapers such as the monarchist *ABC* or the Catholic-conservative *El Debate*. Although airmen had never been noted for their piety in the past, now they were making a political statement of regularly attending Mass in their sky-blue uniforms. To the scandal of his own very Catholic family, Major Hidalgo de Cisneros had taken advantage of the new law allowing divorce and had married Constancia de la Mora, a woman divorced from Manuel Bolín, by coincidence the brother of Luis Bolín who, from London, had organized Franco's important flight from the Canaries to Morocco.

Sometime after returning from Rome, Hidalgo de Cisneros was posted as second in command at Tablada, the military base near Seville, which, he says, was notorious for the reactionary views of its officers. Such was the atmosphere that he was cut dead in the officers' bar and the mess. His reaction was to introduce an intensive rate of training, starting at five every morning and continuing for hours on end, and to insist on strict formation flying. He discovered that the mechanics were discreetly keeping guard and inspecting the plane he flew lest his enemies interfered with it.

In February 1936, the electoral victory of the Popular Front led to the appointment to important posts of senior and trusted officers. One of them was General Miguel Núñez de Prado who was made head of the air force. He immediately posted Hidalgo de Cisneros to Madrid as his personal assistant and trusted subordinate. They and Lieutenant Colonel Luis Riaño, another highly placed officer of progressive Republican views, saw their task as to move politically suspect officers to posts where they could not threaten the Republic. When Santiago Casares Quiroga became Prime Minister of the Popular Front government on 13 May 1936, Hidalgo de Cisneros was appointed his aide-de-camp, but he was ordered to continue his task of moving distrusted commanding officers out of the two main Madrid bases, Cuatro Vientos and Getafe, and instructed to post reliable Republican officers to the other major airfields. Hidalgo de Cisneros and some other loyal officers removed weapons stored in preparation for a coup at the flying school at Alcalá de Henares, some 20 miles east of Madrid. The tension among the officers in the early summer of 1936 was great. Hidalgo de Cisneros was a likely target for hostile officers or for the gunmen who had assassinated a well-known left-wing infantry officer, Captain Faraudo. Hidalgo de Cisneros and others spent sleepless nights watching out for signs of insurrection in the Madrid air bases and organizing vigilance with the use of the Socialist Youth movement and the Communist Party. Such actions were interpreted by conservative officers as preparation for the revolution which they were already anxiously expecting.

During the night of 12–13 July 1936, armed police officers took revenge for the murder of another left-wing officer by arresting and murdering José Calvo Sotelo, leader of the right-wing National Front. Although the uprising had been long planned, the murder by uniformed police of a leading opposition politician probably resolved many officers' doubts. One week later, the garrisons in Spanish

Morocco declared martial law. General Núñez de Prado, head of the air force, made telephone calls from his headquarters in Madrid to all the bases in order to check their loyalty. He was unable to get a reply from Melilla, in the eastern part of the Spanish zone of Morocco, whose commander had already been killed while resisting the insurrection. All the other base commanders, including those who would later rise in rebellion, reported that all was calm. Hidalgo de Cisneros and Núñez de Prado drove personally to the three airfields of Madrid, Cuatro Vientos, Getafe and Barajas, and assured themselves that there would be no insurrection in them.

In Madrid, rebellious army officers had concentrated a number of regiments in the Montaña infantry barracks. On Monday, 20 July 1936, flying low over the Montaña barracks and the artillery barracks at Carabanchel, to the south-west of the capital, loyal bombers dropped their missiles directly into courtyards, covered the barracks with a dense cloud of smoke and dispersed a column of soldiers organized by an infantry general at other military establishments. Combined with artillery fire directed by loyal officers, the insurgents, many of whom were no more than 20-year-old boys undergoing their compulsory military service, surrendered. Loyal aircraft under the direction of Lieutenant Colonel Antonio Camacho, who was officer commanding Getafe air base near Madrid, and Captain Manuel Cascón, formed three columns of troops, added militia from the Socialist Youth and sent up aircraft with bombs and machine guns, which put an end to the insurrection in the nearby artillery barracks. Cascón would pay for his loyalty to the Republic when, at the end of the war, the victors court-martialled and executed him. Camacho, who had distinguished himself in the latter part of the Riff war in Morocco, and had held commands in most of the major airfields of Spain, became Undersecretary for Air during the war. He managed to leave Spain at the end of the war in a British destroyer and was briefly in England. Later, he made his way to Mexico where he died in 1974.[17]

Thus the loyalty of airmen was instrumental in defeating the military insurrection in the capital of Spain, as they were to some extent in Spain's second city, Barcelona, where Lieutenant Colonel Felipe Díaz Sandino assured the loyalty of the air base and ordered his Bréguet XIXs to bomb insurgent barracks.[18] He became Counsellor for Defence of the autonomous government or *Generalitat* of Catalonia. The Russian journalist Koltsov gives a vivid picture of what was probably the passenger lounge in Barcelona's El Prat airfield in early August 1936, with pilots in flying gear lounging on divans, an improvised bar full of men shouting, maps on the walls and, in the middle, grey-haired Colonel Felipe Díaz Sandino trying to impose order and discipline.[19]

The base at Los Alcázares (Murcia), under the command of Major Ortiz, remained loyal and captured the nearby naval aircraft base at San Javier, which had risen in rebellion together with a number of naval officers. The naval aviators had intended to bomb Los Alcázares and to join what they had hoped vainly would be a successful insurrection in the Cartagena naval base.

The insurgents triumph in Morocco, Seville and Majorca

In Morocco, the High Commissioner was an air force officer, Arturo Alvarez Buylla, who telephoned Madrid and told the commander of the air force, General Núñez de Prado, that no officers would speak to him except the commander of the airfield at Tetuán, capital of the Spanish protectorate. This was Major Ricardo de la Puente Bahamonde, a cousin of General Franco on his mother's side. Within hours Alvarez Buylla would be killed, while Franco's cousin would be arrested after resisting the insurgents for some hours. He would be tried by drumhead court martial and shot for 'military rebellion', that is for not obeying the declaration of martial law by the insurgents. The commander of the airfield at Nador in the east of the Protectorate fled for his life into the French zone of Morocco. At the seaplane base of El Atalayón in Melilla on the Mediterranean coast of Morocco, Captain Virgilio Leret was executed for resisting the coup. As a result, there was no point in General Núñez de Prado flying, as he had planned, to Spanish Morocco to forestall a coup, for by the end of the first twenty-four hours of the military uprising Morocco was already lost to the Republic. He would try to block the insurrection in Zaragoza and lose his life in the process. The same fate befell many other loyal army generals in local command all over Spain, while others were imprisoned and later dismissed.[20]

Unfortunately for the Republic, the base at Tablada near Seville was taken over by the insurgents. This would mean that the latter had an airfield to which they would be able to airlift men from the professional forces in Morocco. Fortunately for the historian, a witness later gave an account of what had happened. This was José Macías Ruiz, a senior mechanic in the service of the Spanish civil airline LAPE, which had Douglas DC2 aircraft which could be used as bombers. He recalled the events of the scorching night of Friday–Saturday, 17–18 July 1936. LAPE, which came under the orders of the Minister of War, was ordered to send two DC2s to Tablada to be loaded with bombs, and thence to fly, together with two Fokker VII aircraft which were already at Tablada, to Morocco and bomb Tetuán and Larache, from which news of the military uprising had arrived earlier that day. The DC2s, which were long-range aircraft flying faster than any Spanish fighter, took off from Barajas, today Madrid's civil airport, just after 3 am on Saturday, 18 July, arriving at Tablada after a flight of one hour forty minutes. The Andalusian heat was stifling, even at 4.45 am. The crew prepared to load bombs but many officers asked them not to. Macías was jacking up a wheel on one of the Fokkers, whose tyre had been let down, when Isabel Arranz, who was LAPE representative at Tablada and whose brother, Captain Francisco Arranz, would fly to Germany a week later among a group dispatched to ask for military aid, begged him not to bomb the Tetuán airfield. 'You know my brother is there', Macías remembered her saying. As the mechanics and radio operators who had flown down from Madrid were installing a bombsight and a launching ramp in the DC2s, and piling small 11-kg bombs on to their seats, they heard a shot. Captain Carlos Martínez Vara del Rey, who would become an ace in the Francoist air force, had driven up in his 'Baby' Morris car, alighted, knelt and fired. Anticipating resistance, the crew had brought revolvers from Madrid and returned fire, wounding Vara del Rey, who fled for safety among his companions. However, he had succeeded in putting bullets into both wheels and the crankcase of the starboard engine of the DC2. This led to a fusillade of shots between the crew and the huts where the officers of the base had congregated. One of the mechanics said to Macías, 'Watch out for the bulls!' A herd of fighting bulls traditionally grazed on Tablada airfield, which had a system for scaring them away quickly when an aircraft movement was due. Macías, still in the pants and vest in which he had been working in the sweltering night, hid in a field of maize and in an irrigation ditch. He managed to get away, fled to Seville and warned the provincial governor that the Tablada officers were in rebellion against the government. At 2 pm that afternoon, he returned. The one unharmed DC2 and the Fokkers took off to drop bombs on Larache and Tetuán in Morocco, using the hastily fitted bombsights. The bombing had no effect on Larache and little on the airfield at Tetuán but did kill some inhabitants in the latter city and thus incidentally had the effect on mobilizing the support of the native population for the insurgents. In the meantime, however, insurgent officers took command of Tablada airbase. So, delay, misunderstanding, false assurances of loyalty by the commander and the swift takeover of Seville, known as 'Red' Seville because of its revolutionary populace, by cavalry general Gonzalo Queipo de Llano, led to the loss to the Republic of this important air base.

Franco sent a seaplane to Cádiz, a seaport which had been rapidly taken over by the insurgents, to bring the ex-commander of the air force, the monarchist General Alfredo Kindelán, to Tetuán. Franco now had a DC2 and three military-version three-engined Fokker VIIb aircraft, previously based at Cape Yubi in Spanish Sahara. The latter flew into Tablada on Monday, 20 July, having picked up men at Tetuán. The first carried an officer and nine men from the 5th Battalion of the Legion. They crammed themselves and their equipment into the cabin. The insurgent airmen at Tablada fired on the aircraft, suspecting it had come from one of the Madrid airfields and was about to launch bombs. The legionaries made ready to hit the ground running. Recognizing the characteristic uniforms and caps of the Legion as the men leapt out of the plane, the airmen ceased fire. The legionaries had arrived from Morocco to reinforce the insurgents in

Seville. A few minutes later another aircraft arrived with ten more legionaries, followed later by more. In this way, aircraft made all the difference in allowing the insurgents to overcome resistance to the coup in the working-class districts of 'Red' Seville, who had no way of resisting the skill, training and savagery of a few dozen Moors and legionaries who were rapidly driven around the narrow streets of the working-class districts of San Julián and La Macarena.[21] Besides, with Tablada in their hands, the insurgents had an airfield to which, if they had enough machines, they could airlift troops from Morocco despite any blockade that the Spanish navy could impose.

A similar takeover occurred in Logroño in the north of Spain. Insurgent planning combined with poor preparation by the Republican authorities also led to the loss of Puerto Pollensa, the important seaplane base on Majorca. Ramón Franco, younger brother of the insurgent general and a revolutionary, would be promoted and given air command of the Balearics, much to the anger of many officers who were outraged that a man whom they considered a 'communist' should be given an active command.[22]

How many aircraft for Franco and how many for the Republic?

Around two-thirds of the aircraft in the Spanish air force remained in the hands of the Republic, though in varying states of repair and air-worthiness. Omitting training machines, small planes belonging to regional authorities and state entities, civil airliners and private and sports aircraft, the Francoist insurgents had about one hundred warplanes while the Republic had about twice as many.[23] As for personnel, Major Ignacio Hidalgo de Cisneros, who was soon to command the Republican air force and was in a position to know, claims that about 35 per cent of the five hundred military pilots in Spain remained loyal.[24] One of the rare memoirs by an airman of the Spanish Civil War is by Andrés García Lacalle, who, having gained a pilot's licence from a civilian school only in 1929 and having been a military sergeant pilot only since 1934, rose to command the entire fighter force of the Republican air force. He broadly agrees with General Jesús Salas, whose elder brother was a famous insurgent ace, when he underlines that the Republic had significant superiority in the air and at least equality in number of pilots when the war began.[25]

Nevertheless, a more recent study, based presumably on the post-war court-martial records of all officers who had not been in the insurgent forces but had remained in the Republican part of Spain, suggests that the middle levels of command had been almost completely in favour of the Francoists. As for personnel, at the Madrid bases, though the commanders were loyal, in Group (*Escuadra*) No.1 at Getafe, as well as at Cuatro Vientos, the great majority of the officers were favourable to the uprising. The exception was the sizeable group of *alféreces* or 2nd lieutenants, to which rank senior sergeants had been promoted by a law of December 1935. In Los Alcázares (Murcia) the overwhelming majority of officers were loyal, but in the neighbouring seaplane base of San Javier none were. Most of the officers at El Prat (Barcelona) at least sympathized with the insurgents as did all the naval flyers at the local base. At Tablada (Seville), only one or two officers were loyal. Forty-five were listed post-war as Nationalist sympathizers. All officers in the smaller airfields of Pamplona, Granada and Logroño joined the insurgents.[26] In February 1937, 333 air force officers were officially dismissed by the Republic, though not all of these were necessarily with the insurgents.[27] Many, probably sympathetic to the Francoist cause or worried that they might be thought in some way disloyal to the Republic and fearing arrest or worse, had gone into hiding. In contrast, at air bases where the insurrection had triumphed such as Tablada or in Morocco, if an officer joined his insurgent comrades, a previously suspicious political or ideological stance might be ignored.

Some pilots managed to take off and fly to the other zone. Dramatically, Félix Urtubi, a sergeant-pilot who was stationed in Morocco, took off from Tetuán at 6 am on 25 July with orders to strafe militia columns in southern Spain, having secreted a firearm in his flying boot. Over the Strait of Gibraltar, he turned and fired four times at the lieutenant of a Moorish *regulares* unit who was flying with him as observer and perhaps to keep a check on him. Urtubi told journalists in Madrid:

> We left Tetuán at 0600 hrs today [...] At an altitude of 1000 ft over the Straits of Gibraltar I turned to the lieutenant and shot him four times – in the forehead, in the chest and through the mouth. I didn't give the traitor time to look at me in dismay and cry 'No! No!' [...] I'd rather die than surrender to the traitors to the government.
>
> With very little fuel left I landed at Getafe, and when the officers pointed their guns at me, I put mine to my temple and asked if Madrid was in Republican hands? If not I'd have shot myself rather than surrender.

At that moment, Lacalle drove up in a starter vehicle. Urtubi knew him well and asked:

> 'Lacalle, in whose hands is this airfield?'
>
> 'The Government's, Félix,' replied Lacalle.

He climbed up on the wing. Urtubi showed him the pistol with which he killed his observer. 'I kept one bullet for myself', he said.[28]

If Getafe had been in insurgent hands Urtubi would have been shot, for murder if not for military rebellion, but for the Republic he became a hero.

Urtubi was later shot down but managed to make his way back from enemy territory dressed as a peasant and leading a donkey. On reaching the Republican lines, he was taken for a spy until he could prove his identity. Urtubi was finally killed on 13 September 1936 when, flying a Nieuport against Italian Fiat CR-32 fighters, he ran out of ammunition and, probably badly wounded, was reported to have rammed the Fiat flown by the Italian-American Vincente Patriarca, who recounted Urtubi's heroism to his captors.Another example of a pilot changing sides in flight was that of Ananías Sanjuan. On 10 November 1936, two Republican squadrons were flying above the airfield at Alcalá de Henares when they saw a Junkers-52 landing from a westerly direction, that is from insurgent territory. Sanjuan, who was piloting the German bomber, had been ordered to taxi the Junkers to the pit in insurgent Avila where bombs were loaded into the bays. He decided there and then to take off and flew the 85 miles to the loyal aerodrome of Alcalá de Henares. Later, on 15 December 1936, after flying for months with the insurgents, who were now referring to themselves as *nacionales* or nationalists, Antonio Blanch, a navy flyer, and a fellow airman killed the lieutenant who was at the controls of his Dornier-Wal seaplane and the other two members of the crew. When the Francoists heard, months later, that Blanch had perished when his parachute did not open, they might have considered that justice had been done.

However, the presence among the insurgents of top-class pilots with experience and many flying hours contrasts with the early deaths in combat of a number of similarly experienced and able men in the Republican forces. A check on available sources throws up the names of only sixty-five pre-war air force officers who served in the Republican air force in the war. Even if this figure is doubled, it remains far smaller than the number of pre-war pilots and others whom the insurgents could count on. Thus, Republican squadrons were frequently commanded by a remarkable number of relatively inexperienced pilots of low pre-war rank. While several senior officers remained in the service of the Republic, far more captains and lieutenants, that is experienced flying personnel, joined the insurgents.[29] Perhaps more importantly, the pre-war records of later nationalist squadron and wing commanders demonstrate how much experience of flying and command they had. Out of twenty-seven fighter squadron leaders, twenty-one had been pre-war air force officers, while this was so of only fifteen out of forty-nine equivalents in the Republican air force.[30] Some highly competent and experienced airmen, such as Captains Avertano González Fernández and José Méndez de Iriarte, were both killed flying for the Republic in the first few days, while Captain Manuel Cascón suffered a nervous breakdown, and Major Alejandro Gómez Spencer, according to García Lacalle, was unwilling or prevented from taking part.[31] Major Juan Aboal was 'relegated to secondary posts'.[32] Indeed, both in the Republican army and the air force some officers were distrusted for political reasons and not given active posts, while others, especially in the Republican zone, preferred not to put themselves forward because they did not want to take part in a war against their erstwhile comrades or were uncomfortable with the social revolution and the murders of people of conservative and Catholic views that were taking place in the Republican zone in the first few weeks of the civil war. On the other hand, the incoherent Republican policy of sending individual aircraft on missions led to the deaths of the few experienced pilots left. Andrés García Lacalle, the sergeant-pilot who would in due course become the commander of the Republic's entire fighter wing, recalls that these losses meant that the fighter wing soon had no experienced leaders left.[33]

Both Nationalists and Republicans were clearly in need of aircraft and pilots with war experience. The news of the conflict spread swiftly and it was not long before aircraft and men to fly them were making their way to Spain.

First operations

In the first few weeks, Republican aircraft operating out of the Madrid aerodromes concentrated on shooting up infantry columns approaching the capital over the Sierra de Guadarrama to the north of the city and trying to prevent Francoist aircraft doing the same to the amorphous forces of political militia and the odd units of the army which could be assembled with officers after the defeat of the insurrection in the capital. These small columns of insurgents were hardly the most important strategic objectives at this early period in the civil war. Indeed, a marked feature of the Spanish Civil War was the lack of Republican awareness or interest in identifying the most important targets. Why, for instance, did Republican bombers not concentrate their forces and bomb the ports of southern Spain, now in Franco's hands, to prevent enemy forces landing there from Morocco? Why were there not similar concentrations over Seville, which General Gonzalo Queipo de Llano took with bluster and a few units of well-directed troops, and Zaragoza, once it was known that General Miguel Núñez de Prado's attempt to maintain the garrison's loyalty had failed? If it had been possible to defeat the insurrections in Madrid and Barcelona with a mere handful of planes, would it not have been possible to identify other strategic targets and concentrate attacks on them? Madrid was being approached by Francoist forces from neighbouring provincial capitals to the north and west of the capital. Why, asks Andrés García Lacalle pertinently, were squadrons of Bréguet XIX bombers from the loyal bases around Madrid not sent to fly threateningly over a nearby city such as Avila?[34]

One senses that García Lacalle is hinting that Republican pilots were inhibited by higher authority. This may well be the reason why they did not attack the

Junkers-52 transports which from late July 1936 were carrying men of the professional army from Morocco to Spain. These German aircraft were escorted, which suggests that attacks were expected, by Heinkel-51 fighters which could probably have been tackled by the Nieuport fighters of which the Republic had many more than the Nationalists. However, since the far more powerful Republican navy was ordered not to continue with its blockade after one or two incidents with British warships sent post-haste out of Gibraltar to stop Spanish Republican warships trying to stop and search British merchant shipping in the Strait of Gibraltar, who were suspected of carrying armaments to the rebels, it may well be that the air force of the Republic received similar orders. The fact is that, to say the least, the Republic was indecisive in its handling of its navy and its air force.[35] Both in the Sierra de Guadarrama north of Madrid and in the south and west of Spain as Franco's expeditionary force marched north that summer and autumn from Seville and then east and to Toledo and Madrid, the aircraft of the Republic were used piecemeal and lost while carrying out missions which had no great consequence, thus ignoring the essential military principle of concentration of effort. However, although this criticism, in which the Republican García Lacalle agrees with the Nationalist supporter General Jesús Salas, is justified, it fails to take into account the chaos of Republican Spain, bereft of authority because of the insurrection itself and where commanders, often of limited experience, could not impose the solutions which they would later see, in hindsight, to have been necessary.[36] For example, a Republican column on the wooded slopes and in the valleys of the mountains north of Madrid, with experienced corporals and sergeants, and trusted officers in charge, would soon have learned how to shelter from light bombs dropped from a Francoist Bréguet XIX. But when the undisciplined and untrained militia heard the whistle and the explosion of bombs, in the scornful words of García Lacalle:

> I think that no other aircraft could have done us so much harm as that solitary Bréguet. I don't know how many casualties it caused, but the panic was so great that the mixed multitude of improvised and war tourists did not stop running until they reached Madrid in the double-decker buses of the capital's transport system which had brought them out, [...] but the tremendous clamour which broke out, together with violent demands insisting on protection from the air, intimidated the air force command to such a degree that it issued the following order: 'Maintain a fighter over the Sierra on permanent watch from dawn to dusk.' This order was obeyed to the letter, with the resulting wastage of personnel and planes, for many long summer days, while more important fronts were neglected.'[37]

As a result, the Republican air force wasted its opportunities. By the time the initial chaos was resolved, not only German fighters but also Italian Fiat CR-32s, which could successfully tackle any Republican Bréguet or Nieuport, were flying in Spain. More German and Italian aircraft arrived in September, while, as Franco approached Madrid the Republic was left with one sole Hawker Fury to defend the capital.[38]

Yet concentrated action on strategic targets was hardly possible in the first three months of the Spanish war, because the Republic's air force command could not direct operations beyond the central part of Spain. In Catalonia, the squadrons acted independently but under the authority of the autonomous government or *Generalitat* of Catalonia, in which Colonel Díaz Tendero was in charge of defence, while Republican pilots in the northern strip of loyal territory acted on their own authority. Not until the autumn did circumstances begin to change.

José Giral's government, which had taken over after the resignation of Santiago Casares Quiroga and another very brief administration which tried and failed to stop the military uprising, itself resigned in early September 1936. It was replaced on 4 September by a wide-reaching coalition under the veteran socialist Francisco Largo Caballero, who took the War portfolio but created a Ministry of the Navy and the Air force (*Marina y Aire*) under the other leading Spanish socialist, Indalecio Prieto.

As his Undersecretary for Air, Prieto appointed Colonel Angel Pastor Velasco, who was soon sent to Prague to negotiate the purchase of aircraft and was replaced by Colonel Antonio Camacho, who had commanded the wing based on Getafe, while Ignacio Hidalgo de Cisneros remained in daily charge of operations and Angel Riaño became chief of air staff.[39] Even so, no independent air strategy was worked out, and bombing and reconnaissance missions tended to be carried out on the demand of army or militia units. The dispersal of forces so enraged the pilots that on 27 September, all the available aircraft at Getafe took off together and drove off nine Nationalist fighters. They were duly reprimanded. As a point of comparison, on 23 August 1936, the Nationalists carried out their first large-scale bombing raid en masse on the major Republican air base at Getafe. Loading 250-kg and 50-kg bombs, they succeeded in destroying nineteen enemy machines standing unprotected on the ground. Republican anti-aircraft defences failed to score any hits. On 26 October, a similar raid destroyed five machines on the ground at Barajas.[40]

Nevertheless, despite these failures, the team of Largo Caballero, Prieto and Camacho began the process of negotiating the acquisition of material, fuel and parts; of training new pilots, observers, gunners and bomb aimers; and reorganizing the aircraft industry to turn out new planes, thus allowing the Republic to continue fighting.

Chapter 2
Off to Spain: The Germans arrive

Speed and efficiency; Germany sends transport planes

By Tuesday, 21 July, after General Sanjurjo's fatal crash in Portugal and the failure of military uprisings in Madrid and Barcelona, it was obvious that the entire enterprise to overthrow the Republic would collapse unless Franco's Moorish troops and legionaries could be shipped over to the Peninsula in significant numbers. By this time, however, the majority of the Spanish navy was either in or rapidly approaching the Strait of Gibraltar. Its mission was to blockade Franco's professional forces, impeding them from crossing from Morocco to Spain. However, the Spanish naval officers had planned, if not to assist actively, not to do anything to hamper the army's insurrection. Radio security aboard was very lax and messages were sent back and forth by Benjamín Balboa, a senior radio warrant officer of pronounced Republican sympathies, from the central radio station in Madrid to the ships, warning the crews of what the officers were planning. The crews of most of the ships mutinied in their turn against the officers, and several hundred of the latter were later murdered by extremist sailors while awaiting court martial.[1]

This put Franco in a quandary. The few ships at his disposal were small and no match for Spain's modern destroyers. The solution would have to be an airlift of men and equipment, but for this Franco had to have control of the air as well as a fleet of transport aircraft with the capacity to move more than a few men at a time. He decided to appeal to Germany, specifically to the German military attaché in Lisbon, General Kühlental.

While there is no indication that the Nazi government encouraged or even had foreknowledge of the Spanish military coup, there had been many connections between the two countries in relation to possible purchases of armaments.[2]

Franco took advice from Lieutenant Colonel Beigbeder, who had been Spanish military attaché in Berlin and was now head of the Department of Native Affairs in Spanish Morocco. Beigbeder used the facilities of the German consulate in Tangiers to telegraph the German Foreign Ministry in Berlin and ask them to instruct the attaché in Paris and Lisbon to ask his superiors to send aircraft:

> Lieutenant-Colonel Beigbeder has asked me to forward the following very secret dispatch: For Military Attaché General Kühlental: General Franco and Lieutenant-Colonel Beigbeder send greetings to their friend, the honourable General Kühlental, inform him of the new Nationalist Spanish Government, and request that he send ten troop-transport planes with maximum seating capacity, through private German firms.
> Transport by air with German crews to any airfield in Spanish Morocco. The contract will be signed afterwards. Very urgent! On the word of General Franco and Spain!
> For the Consul:
>
> Wegener[3]

The cable was received early on Thursday, 23 July. After consulting the German embassies in Paris, Madrid and London, the German Foreign Ministry, known from the street it occupied as the Wilhelmstrasse, rejected Franco's appeal. Germany had normal diplomatic relations with Republican Spain. The Wilhelmstrasse followed the rules. One did not provide arms and men to help a rebel against a friendly government.

Nevertheless, where one door shut another opened, and this rebuff to Franco's appeal was replaced by two opportunities. The first was already there. In the Canaries, the local representative of the German airline Lufthansa, ex-naval lieutenant Otto Bertram, who was in charge of arrangements for possible German naval requirements in the area and in direct communication with the German navy and the *Auslandssorganisation* or overseas direction of the Nazi Party, had been asked on 15 July by General Luis Orgaz, a long-suspected military conspirator whom the Spanish government had exiled to the Canaries, if he could provide him with an aircraft to fly Franco to Spain, perhaps as a backup if the expected British plane did not arrive from Croydon. As it happened, the De Havilland DH-89 Dragon Rapide, chartered by the journalist Luis Bolín and piloted by Captain Bebb, did land soon after. Bertram told Orgaz that he did not have the level of authority to loan him an aircraft, and in any case he suspected that some political motive was afoot and realized that he ought not to become involved. His suspicions were of course justified, for on Saturday, 18 July 1936 at 5.30 in the morning, General Franco declared martial law in the Canaries and radioed a call for insurrection to all the garrisons of Spain.

In Las Palmas on Grand Canary, the following Monday, 20 July, at 11.37 am, 34-year-old Lufthansa captain Alfred Henke landed his Junkers-52/3m aircraft, named 'Max von Müller', at Gando aerodrome with his regular postal flight from Bathurst (the Gambia) via Villa Cisneros in Spanish Sahara, along one of the air mail routes which Lufthansa, the most extensive airline in Europe, was pioneering. By this time, most of Spain was fighting; the military insurrection had been defeated in Barcelona and was struggling in Madrid. Its putative leader, ex-General Sanjurjo, was soon to die when his Puss Moth crashed in Portugal. Orgaz told the protesting Bertram that he had to requisition the Lufthansa Junkers-52 but that he would deposit a sum in a local bank to cover any damage to the aircraft. Against the loud objections of Bertram, whom the Spanish military even briefly arrested, the insurgent officers said they would compel Henke to fly the Junkers-52 from Las Palmas to Tetuán. Bertram cabled the news to Berlin and Bathurst and advised that it would be best to suspend all planned flights in the area. Late on Tuesday, 21 July, 'Max von Müller' left Las Palmas for Tetuán, arriving very early the following morning. The Lufthansa Junkers was to have a vitally important role to play, for just as Olley Air Service's Dragon Rapide would fly into history as the machine which had taken Franco from the Canary Islands to Morocco to lead the Spanish army over to Spain and vanquish 'Bolshevism' there, so 'Max von Müller', registered as D-APOK, would play a dramatic part in the next few days and in due course lead to the massive German air intervention which would be one of the major causes of Franco's victory over the Spanish Republic.

Chance now took another hand. The Wilhelmstrasse had rejected Franco's appeal for aid, but Hitler and the Nazis did not play by the same rules as the diplomats. A second opening presented itself. In Tetuán, capital of Spanish Morocco, there was a small German business colony, one of whose members was Johannes Bernhardt, a businessman with important connections with the Spanish military, and another was Adolf P. Langenheim, a 64-year-old mining engineer who had spent most of his career in Morocco and was now the *Ortsgruppenleiter* or local organizer of the thirty-three Nazi members of the colony. As such, he reported to the *ausslandsorganisation* of the Nazi Party, which supervised party members abroad.

That same morning, Thursday, 23 July 1936, Bernhardt offered to fly to Berlin, bypass the bureaucracy of the Wilhelmstrasse and seek help through the Nazi Party. Here was Franco's opportunity and perhaps his only chance. A German civilian aircraft stood ready on Sania Ramel airfield at Tetuán. The local Nazi leader and a German businessman, both well known to the Francoists, were willing to use their connections with the Nazi Party to serve the military uprising, already known as the Glorious National Movement to free Spain from Bolshevism. Franco seized his chance.

So, at 5.30 pm on Thursday, 23 July 1936, Langenheim, Bernhardt, a rather unwilling Henke who not unnaturally feared for his job if his employers disapproved of his piloting a Lufthansa aircraft to help what he insisted was a *Räubergeneral* ('bandit general') and Captain Arranz, a Spanish air force officer, boarded 'Max von Müller'. The Ju-52 bumped uncomfortably down the grass runway, baked hard by the summer heat. Bucketing in the fierce air currents, they flew across the Strait of Gibraltar to Seville, which had been captured by the Nationalist insurgents, and, after a small repair the next day at the Tablada air force base, took off on a long and tiring flight out to the coast at Valencia, up Spain's Mediterranean coast, landing at Marseilles and Stuttgart, and on to Berlin, where they touched down as the summer dusk fell. Aware of the flight, perhaps because Henke had communicated with the local German agents of Lufthansa, the German authorities, concerned to keep the arrival of Franco's emissaries secret, had ordered the pilot to land at the military aerodrome of Gatow rather than the civil airport of Tempelhof.

The next morning, Saturday, 25 July 1936, through the agency of Friedhelm Burbach, head of the *Auslandssorganisation* or *AO* for Spain, Ernst Wilhelm Böhle, the overall head of the *AO*, was persuaded to telephone their mutual school friend Alfred Hess, brother of Rudolf Hess, the deputy Fuehrer, who would himself take a dramatic flight to Scotland during the Second World War. Alfred Hess grasped the significance of the mission of the two Germans from the minute community in Morocco and rang his brother, who was on holiday at the spa of Bad Kissingen. Rudolf Hess, the deputy Fuehrer, realized the importance of the mission and picked up the telephone to speak to Adolf Hitler. The Fuehrer agreed to meet the emissaries. That afternoon, Saturday, 25 July 1936, only a week after Franco had declared martial law in the Canary Islands, the party flew to Nuremberg in an aircraft provided by the director of Lufthansa and drove on to Bayreuth where the Fuehrer was attending the Wagner festival.[4]

One cannot but be struck by the fact that everybody involved had answered telephone calls, despite it being Saturday, for later that same day, just twenty-four hours after arriving in Germany, the two businessmen from the tiny German expatriate colony in Spanish Morocco found themselves in front of the Fuehrer himself and his entourage, among them Field Marshal von Blomberg, head of the army, and Hermann Goering, head of the Luftwaffe, the new German air force which only the previous year had emerged from the secrecy which it had had to observe because of the terms of the Treaty of Versailles. When Hitler returned from the opera house, he ordered Franco's letter to be read and translated. The Spanish general had explained his aims and was asking for ten transports, six fighters, anti-aircraft guns and sundry other war material. Hitler formulated his thoughts and within two hours had made his decision, quashing the opposition of some of the members of his entourage and ignoring the advice proffered by Wilhelmstrasse officials earlier that day. It would be his first adventure abroad. He had probably heard of Franco, or his advisers might have reminded him that the general had been chief of staff during the military repression of the Spanish left-wing uprising in the Asturias mining basin in October 1934. Hitler ordered Hermann Goering, his air chief, to make arrangements to provide the almost unknown Franco, 1,500 miles away, with his needs. Goering summoned General Milch, his Undersecretary, a man of extraordinary ability who had developed Lufthansa as the leader in European civil aviation, while Rear Admiral Lindau was woken in his bed in Hamburg and told that he would be picked up by plane early next morning and brought to Bayreuth to receive orders. By the next evening, Sunday, 26 July, Lindau was back in Hamburg with orders to prepare dock facilities for loading a cargo ship with equipment and an expeditionary force, while Milch had returned to Berlin, had summoned a group of Luftwaffe officers and had ordered General Helmuth Wilberg, one of the pioneers of military aviation, organizer of pre-Nazi epoch clandestine training in the Soviet Union for the still-secret Luftwaffe and a leading theorist of the use of the air arm, to create a special staff – to be known as *Sonderstab W* – to organize the dispatch to Franco of twenty Ju-52/3m transport aircraft with fighter escorts and the necessary ground backup.

They would use Ju-52 planes already in Luftwaffe bomber squadrons or ready for delivery to Lufthansa and form a commercial company to hide official German intervention in the Spanish conflict. The operation would be called *Unternehmen 'Feuerzauber'* or 'Operation Magic Fire', recalling the 'Magic Fire' with which Wotan had surrounded the sleeping Brunnhilde in the opera which Hitler had watched in Bayreuth earlier that evening.

Few such major political decisions can have been taken so quickly as the Nazi leader's resolve to aid Franco. But Hitler's apparently risky move was inspired by his extraordinarily acute political judgement. He had a number of reasons for helping Franco. Considering that his first decision was merely to supply transport machines to airlift Spanish insurgent troops over the Strait of Gibraltar, it is unlikely that he had thought of exercising the young Luftwaffe aggressively, though this consideration may have been in the mind of Luftwaffe overall head Hermann Goering that Saturday night and the next day as he briefed Generals Milch and Wilberg. The logistics of an airlift carried out at 1,500 miles' distance with probably insufficient facilities were in themselves an excellent training and rehearsal for a task which the Luftwaffe might one day be called on to perform. Nor were the economic considerations, especially the potential supply of raw materials from mineral-rich Spain, later so important, uppermost at the time in the Fuehrer's mind. Hitler would have thought first of the strategic value of having a Spanish regime in power hostile to France's Popular Front government. He was probably aware how important neutral Spain had been as an espionage centre and a supply base for German U-boats during the 1914–18 war. Furthermore, Hitler probably knew that Mussolini, the Italian Fascist dictator, whom the German leader considered a model to imitate, was pondering whether to help Franco. Perhaps the German leader thought that he could not afford to let Italy take the lead. Over all, nevertheless, loomed Hitler's hatred of communism and the Soviet Union, which he believed, in common with much right-wing opinion in Western Europe, was infiltrating the Continent through the policy of the Popular Front and the apparent union of communists, left-wing and middle-of-the road parties, together with the communist abandon of revolution, replaced by support for parliamentary, democratic regimes. Nevertheless, one might well wonder whether Hitler would have troubled himself about remote Spain had he not been visited on 25 July by the two German businessmen from the tiny expatriate Nazi group in Spanish Morocco, who had thought of using the air route to reach the Fuehrer quickly through the Nazi Party when Franco's appeal for help had already been turned down by the Wilhelmstrasse.

During the next two days, Captain Henke's Ju-52 was serviced and its inner fittings were stripped out to make room for hundreds of cans of extra fuel, for Henke was ordered to fly back to Morocco non-stop. There was hardly room for the passengers, who must have endured the stink of aviation fuel as they tried to doze through several uncomfortable hours. They took off on the night of 27–28 July, reaching Tetuán at one o'clock the next afternoon. Because of his experience, Henke, though a Lufthansa pilot and not in the Luftwaffe, was retained to fly one of the aircraft which were to be supplied. Erwin Jaenecke, the chief of staff of *Sonderstab W*, wrote later: 'In Spain people like Henke are more useful than ten bombers.'[5] He probably meant that Henke could speak Spanish and was used to what the meticulous and punctual Germans would have considered the irritating lack of order and discipline of the Spaniards. Not till 1938 did Henke rejoin Lufthansa when on 10–11 August he flew the four-engined Focke-Wulfe 200 airliner four thousand miles non-stop from Berlin to New York in a world record time of twenty-five hours. His co-pilot was Rudolf von Moreau, who had flown with Henke in Spain.

While Henke saw to the removal of German insignia from his aircraft, and then, rather unsafely because he must have been exhausted, flew his first group of Moorish troops over the Strait of Gibraltar to Seville, Langenheim and Bernhardt brought the delighted and relieved Franco the news that twenty aeroplanes, rather than the ten he had first requested, were coming. Franco would receive anti-aircraft guns and Heinkel-51B fighters to escort the Junkers against the inferior Republican Bréguet XIX and Nieuport-52 fighters that might try to attack them as they airlifted Franco's forces across the Strait of Gibraltar.

Over the next few days, nine more Junkers flew to Morocco via Switzerland and along the Italian, French and Spanish coasts. On 31 July at midnight, to ensure maximum secrecy, the rest of the Junkers, brought from the factory at Dessau, and the Heinkels, together with the anti-aircraft guns, left Hamburg, packed in 773 crates, on the Woermann Line's merchant ship *Usaramo*, to arrive, escorted from the Portuguese coast by German warships under Admiral Carls, at the southern Spanish port of Cádiz on 6 August.

One Ju-52 took off from Germany to fly to Spain but landed in error at Barajas (now Madrid's civil airport) mistaking it for insurgent Seville. One of the pilots of another German aircraft which had arrived in Madrid to evacuate civilians told the crew that they were in the Republican zone. The pilot revved up his engines and took off in haste. Out of fuel, he landed again at Azuaga in the province of Badajoz in western Spain, by which time the Spanish government had been alerted and ordered the plane to be impounded and flown back to Madrid where the German ambassador, who could not betray his knowledge that his country was helping the Franco insurgents, protested that the aircraft had arrived to repatriate German civilians, even though it had been fitted out for military purposes. Sometime later, the Ju-52 was destroyed in an air raid.[6]

Maintaining an expeditionary force in Spain, 1,500 miles away, was a difficult operation but one which the Luftwaffe was psychologically prepared to undertake. For, although Germany had been forbidden to have an air force by the treaty of Versailles of 1919, it had nevertheless been reconstructing its air force since the very end of the First World War. A comprehensive list of all surviving air force personnel had been established by the *Ring Deutsche Flieger*, an organization of aircrew. The existence of the Ring was known and evidently tolerated by the victorious Allies, for the prestigious British *Royal United Services Review* reported in mid-1922:

> The fact that, as far as possible, every link that forms the 'ring' is to consist of the combatants who served together in one squadron or unit, coupled with the violent invective used by the chief speaker at its opening meeting, shows that the future activities of this society will probably be directed towards keeping in being, though in a disguised form, an air force imbued with the ideals of the old one [...] strong financial and industrial interests, if not the government, are offering their support.[7]

Within the small army which Germany had been allowed to maintain after 1918, a small group of officers, innocently titled the *Truppenamt* or 'Troops Office' including Captain Helmuth Wilberg, who as a general would be in ultimate charge of *Sonderstab W*, which organized the expedition to Spain, had studied in extreme and disciplined detail the lessons in the organization of squadrons, combat tactics and technical developments to be drawn from the First World War. Another of the air force officers who participated in these study groups was Hugo Sperrle, who had been officer commanding the air force of an entire German army. Sperrle helped to organize the secret air force of the Weimar Republic's army, the *Reichswehr.* The Treaty of Versailles had prohibited the development of German civil flying for only six months and by the 1930s German airlines were flying more miles than their French, British and Italian equivalents combined, thus giving Germany experience in long-distance flying, multi-engine aircraft, night flying, navigation and the use of sophisticated instruments, which would be invaluable in war. The German army, from which so many air force officers would be drawn, introduced a highly rigorous process of selection, which demanded advanced educational standards, and an intellectual seriousness which commanding officers were required to instil. All officers had to be able to pass the level of examinations which would, previously, have selected only an elite to attend the staff college. Their foreign language ability and their technical knowledge had to be first-class. The old general staff no longer existed, but officers who aspired to that level of appointment could do so with a degree from a technical university. Wolfram Von Richthofen, for example, who would be chief of staff of the Condor Legion in Spain, completed an engineering degree in 1925 at the *Technische Hochschule* at Hanover and was immediately assigned to the secret army staff in Berlin. In the late 1920s, he was attaché in Italy and as such reported in depth on Giulio Douhet's revolutionary ideas of winning war outright by decisive bombing of the enemy's cities. By 1920, some experienced German air force officers were attending the fifty-seven committees which were revising knowledge of everything to do with war in the air. Thoroughness was a German characteristic. Despite the Versailles ban on an air force, during the Weimar Republic which preceded the Nazi era, German airmen, like their army counterparts, had continued their traditional zealous study of the science of war. Close scrutiny of the experience of the First World War, together with detailed training programmes and tactical exercises, followed by full reporting and constant self-criticism, maintained the new Luftwaffe at the highest level of professional and disciplined efficiency.[8]

In 1927, in order to avoid the restrictions placed upon the German armed forces, Germany reached an agreement with the USSR to establish a training and experimental flying centre at Lipetsk, some 220 miles south-east of Moscow, which became a school for advanced air tactics, not closed until the Nazis ended cooperation with the USSR. Experiment and training at Lipetsk led to the writing of constantly reviewed manuals covering all aspects of war in the air.[9] Lipetsk allowed the Germans to practise live-fire exercises, fighter versus bomber tactics and dive-bombing. Thus the future Luftwaffe, whose existence was not even admitted until 1935, possessed a fighting doctrine buttressed by close study, intellectual analysis and practical application. The lessons of the First World War were absorbed, but just as in the parallel conclusions in the army, the German experts did not assume that the next war would be like the 1914–18 conflict, a stalemate based on trench warfare, but a war of movement, *blitzkrieg* or 'lightning war' in which air power would have a vital role, though, unlike Italy's strong belief in the theory of Giulio Douhet that bombers on their own could crush the enemy will to resist by destroying his major cities, the incipient Luftwaffe closely debated the combined roles that bombers and fighters would play in the future air war. The later General Wilberg, who had ultimate charge of the German expedition to

Spain, had been the adviser to Hans Von Seekt, creator of the post-First World War *Reichwehr*. As early as the end of 1919, Wilberg had created committees to examine major aspects of war in the air against the background of what had been learnt and what were thought to be the technical developments about to come into being.[10] Luftwaffe Regulation 16, 'The Conduct of the Aerial War', written by Wilberg, insisted that success in war depended on gaining air superiority, but the air force had many and varied roles to play. The air force which could successfully carry out such varied tasks would be the winner. Now, in Spain, these doctrines would be seen in practice and evaluated meticulously.[11]

When the Nazis came to power at the end of January 1933, they resolved that Germany should quickly become a major air power by carrying out an intensive four-year programme of airplane construction under Hermann Goering, who was made Minister for the Air as well as supreme commander of the Luftwaffe. This would be the start of massive expansion and a new generation of warplanes. Germany walked out of international negotiations on arms reduction, recreated the army, introduced conscription, rebuilt its navy, reaching agreement with Britain in the Anglo-German Naval treaty of June 1935, and began its aggressive international behaviour by remilitarizing the Rhineland in March 1936. Concretely, between February and August 1936, the Luftwaffe trebled the number of its squadrons. Germany's frontline air strength when Hitler resolved to aid Franco on 25–26 July 1936 was 1,833 aircraft, including 450 Junkers-52 transports which could be used as bombers.[12]

The personnel of Operation 'Magic Fire', consisting of twenty-five officers and sixty-six non-commissioned officers, other ranks, civilian technicians, radio specialists and mechanics, and a medical officer, were rapidly selected from Luftwaffe stations.[13] The commander was Major Alexander Von Scheele, who had lived in South America and spoke Spanish. He had orders not to let his planes fly combat missions, but his fighters were there to protect the Junkers transports and he was possibly ordered to use his judgement about how to do this.

Within eight days of Henke's return to Morocco, the three-engine Ju-52 transports, with their cabin interiors for seventeen passengers stripped out, had each carried forty-five Moroccan and Legion troops in about forty flights to Seville, each of which took one hour and forty minutes, and, as soon as it was made ready, to an airfield at sherry capital Jerez de la Frontera, 50 miles closer. In the week of 10–16 August, nearly eight tons of armaments and 2,853 men or four hundred per day, sitting on the floor with their rifles between their knees, had been flown, making this the first military airlift in history. The legionaries and Moors waited, camped out on the airfield at Tetuán. Conditions were difficult. The August heat was unbearable. Aircraft flew all day, carrying men to the Peninsula, so they had to be serviced at night in the light of headlights and cooking fires. The shortage of adequate equipment meant that aircraft had to be refuelled with hand pumps. When the runway at Jerez de la Frontera was ready, sherry pumps were used but they were slow and took a minute to pump three litres of fuel.[14] German meticulousness had to admit that the Spanish gift for improvisation or *chapuza* was amazingly useful at times.

One month later, by 28 August 1936, Germany had despatched twenty-six bombers, fifteen fighters, twenty anti-aircraft guns, fifty machine guns and eight thousand rifles to Franco's insurgents or, as his defenders insisted, Nationalists. From then onward, the emphasis in the airlift was on war material rather than men, peaking in the week of 14–20 September 1936 at 69.5 tons. The German airlift of Franco's forces from Morocco to Spain came to an end on 11 October, having transported 13,962 men and 270 tons of war material including 36 pieces of artillery in 868 flights. Only one aircraft and two men had been lost, a feat which underlines the professionalism and discipline of the airmen and particularly of the maintenance crews, mostly Spanish, who serviced and refuelled the aircraft without many of the proper tools and in conditions of extreme heat at Seville's Tablada base where day temperatures in July and August can reach 50 degrees.

German aircraft and crews had flown enough of the professional Spanish army from Morocco over the Strait of Gibraltar to enable control by Franco's expeditionary force of most of Western Andalusia and had ensured that the Republican naval blockade failed, thus clearing the passage for Franco's troopships and allowing the insurgent general to begin his march on Madrid. Adolf Hitler said later, 'Franco ought to erect a monument to the glory of the Ju-52. It is the aircraft which the Spanish Revolution has to thank for its victory.'[15]

Despite German efforts to conceal the nature and purpose of the arrival of men and aircraft, the secret could not be totally kept. British Intelligence services intercepted a message from the Spanish consul at the international city of Tangiers, about 27 miles from Tetuán, to his government in Madrid reporting that German aircraft were airlifting troops to the Peninsula.[16] The British vice-consul at Tetuán reported that by 5 August over twenty 'large aircraft' were at the local airfield.[17] In Hamburg, dockers were threatened by the Gestapo if they mentioned anything about the shipments but there is some evidence that surviving anti-Nazi organizations on the Hamburg waterfront did get some information out.[18] The news soon broke in the British press, for the journalist Arthur Koestler spotted German airmen at lunch in the luxury Hotel Cristina, on the banks of the Guadalquivir River in Seville. He had been sent by the exiled German Communist Party to investigate rumours about German aircraft and pilots aiding Franco. He writes in his *Spanish Testament*:

> There were four of these gentlemen in the Hotel Cristina in Seville at about lunchtime on August 28th, 1936. The four pilots were sitting at a table, drinking sherry. Their uniforms consisted of the white overall worn by Spanish

> airmen; on their chests were two embroidered wings with a small swastika in a circle.[19]

The badge was that of Hispano-Marroquí de Transportes or HISMA, the front set-up to disguise the German aircraft as a commercial arrangement so that as yet there was no proof that the pilots or the planes belonged to the Luftwaffe. Koestler may have been mistaken about the white uniform. Some of the Germans wore the whites which had been the dress of the security staff at the June Olympics. Perhaps they picked the clothes up at the Döberitz Olympic village, where they had assembled after volunteering for Spain.[20]

Koestler had got into Nationalist Spain by pretending to be a reporter for a right-wing Hungarian newspaper. In the Hotel Cristina in Seville, he was recognized as a communist by a German journalist sitting with the pilots. Fearing arrest, Koestler rushed out of the hotel and within an hour was speeding in a taxi towards Gibraltar. Back in London, he published his account in the *News Chronicle* of 1 September 1936. When he returned to Spain in 1937, he was captured by the Nationalists. His record put him in gaol for several weeks under, as he thought, sentence of death, until a campaign by the *News Chronicle* and in Parliament secured his exchange with a Francoist pilot.[21]

Thus by September 1936 at the latest, it was no secret that Nazi Germany was aiding Franco, despite Germany's decision to accept the pan-European Non-Intervention agreement, accepted by all countries in August, not to supply war material to either side in the Spanish conflict

Luftwaffe lieutenant Hannes Trautloft was summoned by his commanding officer and was invited to volunteer for a secret mission. Trautloft had had experience of this level of secrecy because he had been sent for training to the Soviet Union during the Weimar Republic as part of the process of recreating the German air force despite the prohibition imposed by the Treaty of Versailles of 1919. Within two hours, Trautloft was on his way from his base near Cologne to Döberitz, today a nature park, but in August 1936 recently an important military base, whose accommodation had been used only a month before as the 'Olympic Village' for the athletes attending the 1936 Games in Berlin. Here he reported on 29 July to be given his instructions and travel documents. Only unmarried men with outstanding records were selected. Many came from a high social class and possessed impressive academic qualifications. They had grown up during the Weimar Republic and were enthusiastic about flying which had given them careers during the Depression of 1929–32. They were the first members of the new air force, the Luftwaffe, and considered themselves the elite of the new Nazi Germany as it emerged from the disgrace of defeat in 1918 and the harsh treatment imposed by the victors. Now these young men would be the first Germans to take part in a modern air war. The Italians had done so the previous year, but against a primitive people, the Abyssinians, who had no chance of resistance. In Spain the Germans would once more fight against Europeans, this time the Spanish Republic under its Popular Front government, supported, as Nazi propaganda proclaimed, by the Soviet Union and the French Popular Front, Germany's ultimate foes.

The Luftwaffe men were ordered to resign their commissions and were placed on the Reserve. If they became prisoners, they could insist that they were not actually in the Luftwaffe. In Spain, they would have higher pay and promotion. One could save perhaps seven thousand marks in a few months, a huge sum for the working-class men who were 'volunteered' for the less glamorous roles as fitters and mechanics. Service in Spain was also good for one's seniority.[22]

Strict secrecy had to be maintained. Adolf Galland, a fighter ace of the Second World War, recalled later, 'One or two of our comrades vanished suddenly into thin air, without our having heard anything about their transfer orders [...] after six months they returned, sunburnt and in high spirits.'[23] However, they were ordered to say nothing about their time in Spain. When they wrote home, they had to address their letters to 'Max Winkler, Berlin SW68'. Their families were threatened with drastic consequences if they betrayed their presence in Spain. The men were ostensibly 'away working' or 'in the Foreign Legion'. Each man was provided with civilian clothes and a suitcase, for the expedition to Spain was to be disguised as the *Reisegesellschaft Union* or the Union Travel Company, a tourist group created for the *Kraft durch Fried* ('Strength through Joy') movement, the national Nazi holiday organization. Great attempts were made to keep German involvement secret. The huge packing cases in which equipment was dispatched were labelled as furniture removal containers. The men did not know that they were going to Spain and were told merely by General Wilberg that they were going to save a people in great difficulties from Bolshevism. Generals Milch and Wilberg inspected the expedition on 31 July before it left Berlin at 11 am for Hamburg to board the *Usaramo*, the cargo ship which was to take them to Spain. The voyage was calm. The ship had cabins for the officers, a bar and good food. The men sunbathed on deck. The next expedition, which left in November, was less lucky. Later winter voyages were plagued with bad weather and seasickness.

During the voyage of the *Usaramo*, Major Von Scheele told the men that they were bound for Spain. Their task was to help the Spaniards to save their country from revolutionary Bolshevism. He told them about Spain, which he knew, warning them to be tactful and not to behave like bulls in a china shop. At the same time, the mimeographed on-board magazine produced by the Signals section filled the heads of the volunteers with a mixture of images of Red atrocities, of which there had been no shortage, combined with an undertow of eroticism about the charms of Spanish women and warnings about the jealousy of their menfolk. They docked at Cádiz in the extreme south of Spain. Later voyages arrived at Vigo in the north-west, but for the first few months all Germans were concentrated first in Seville, the Andalusian capital. Here, palm trees, sumptuous buildings, many in the pseudo-oriental style introduced during the period of the Moroccan wars in the 1920s, and the genuine oriental buildings dating from the Moorish epoch of the Middle Ages, extremely hot summer weather reaching 45 degrees centigrade or more, air heavily perfumed by the rich vegetation, all combined to make the Germans feel a profound sense of oriental exoticism.[24]

They were, however, in Spain for a specific purpose, which at the beginning was to carry out an airlift of Franco's legionaries and Moors from Morocco to Andalusia. The Junkers took off from the airfield at Tetuán, which had only what for the Germans must have been primitive facilities, and flew with little in the way of navigation aids. Both sides in the civil war had to rely on Michelin road maps. The Junkers normally carried seventeen passengers. Flying forty-five airsick Moroccan troops and fearsome legionaries, who insisted on smoking, over the Strait of Gibraltar was a mixed experience for the German pilots. The Nazi sense of racial superiority with which they were imbued made them look askance at what they considered the Moorish 'semi-savages', with strange habits and probably singing and shouting in Arabic, whom they had to carry for up to ten hours each day. An anonymous journalist later wrote that, in the heat of summer, the Moorish troops brought a great deal of impedimenta with them, cushions, tambourines, teapots, tea and sometimes animals to be slaughtered for food.[25]

The expedition included six fighter pilots who were to show Spanish pilots how to fly the He-51 fighter escorts. They had a frustrating wait at Tablada base near Seville while the huge cases arrived by train from the port with the different parts of their fighters. In torrid heat, the mechanics assembled the machines. The excessive speed with which the expedition had been put together meant that even German efficiency had not managed to ensure that the right tools were available, and some basic facilities were not to be found. The Germans worked with Spanish mechanics, suffering the usual problems of mutual understanding because interpreters had not been provided. Von Scheele's report to *Sonderstab W* was particularly critical of Spanish lack of punctuality in providing food. The Germans wanted their main meal at noon precisely, while Spanish mealtimes could be delayed for hours. Occasionally, German mechanics fainted in the searing heat while hands were burned on tools which had been left in the sun.

Within a few days, the Heinkels had been assembled and were ready to fly. Lieutenant Herwig Knüppel recalled that he was not impressed by the discipline. The aircraft stood in the open. Their orders were not to engage in combat lest they be shot down and their presence become officially known to the recently established Non-Intervention Committee. So they began to train Spanish pilots including Captains Luis Rambaud and Joaquín García Morato, and Lieutenant Julio Salvador. These and other formed what they called the 'Rambaud Squadron'.

In early August, pilots of the Junkers-52 transports reported that, as they flew from Morocco to Seville, they had been fired on by the anti-aircraft guns of the Spanish battleship *Jaime Primero*. Although they had been ordered not to engage in combat, von Scheele now interpreted instructions in the sense of using his initiative. This is another example of the extent of improvisation and freedom of action which the expedition to Spain enjoyed, at least in its early days. In this case, given that the aircraft had been attacked, Von Scheele judged that if he ordered a response, he would be acting within the meaning of his orders. Consequently, two Ju-52s were taken off transport duties and fitted with elementary bomb racks.

Intelligence, possibly from the numerous German warships in the neighbourhood, reported that the *Jaime Primero* was berthed at Málaga. At 4 am on 13 August, the two Junkers took off from Seville, flown by Lufthansa captain Henke and Lieutenant Rudolf Baron von Moreau, who was the squadron commander of the Junkers, with a junior Spanish officer and German lieutenant Max Count Hoyos aboard. They flew south and crossed the coast at Estepona, now a tourist centre on the Costa del Sol, and continued eastward along the coast towards Málaga, which Henke reached at 5.30 am though Von Moreau lost his bearings in cloud. The *Jaime Primero* was not tied up at a berth but the Spanish naval officer spotted it anchored in the bay. On the second pass at a height of 1,500 feet, Henke's Junkers dropped three bombs, scoring a hit on the bridge, flying off when the battleship's anti-aircraft guns opened fire. Three of the battleship's crew were wounded, one man was killed and two disappeared, perhaps blown overboard.

Henke and Moreau continued their special operations on 23 August and again a few days later, when they dropped food, medical supplies and letters to the 1,300 insurgent *Guardias Civiles*, the militarized Spanish police force, who were under siege from Spanish government militias in the fortress or Alcázar of Toledo in central Spain.

The Germans had been ordered to instruct Spanish pilots how to fly the He-51 fighters that had arrived in the *Usaramo*. The Spaniards, however, insisted on flying them themselves. Two ace Nationalist pilots, Joaquín García Morato, with thirteen years' experience and 1,860 flying hours in his logbook, an instructor in instrument flying, who excelled in acrobatics and with plentiful experience in the Riff War, together with Julio Salvador Díaz-Benjumea, with 550 flying hours to his credit when the military insurrection began, flew the He-51 successfully, shooting down two Republican Nieuport-52 fighters, a Bréguet XIX biplane and a Potez 54 bomber, but most of their comrades found the Heinkel hard to handle and damaged three of them. Although General Kindelán, head of the Nationalist air force, wanted Spaniards to fly the Heinkels, Von Scheele refused, taking the remaining He-51s back and, despite his orders, allowing the German fighter pilots to participate in combat missions. Hannes Trautloft's memoirs insist that the Germans were determined to fly their planes rather than let the Spaniards do so. Finally, a fourth aircraft was constructed with pieces salvaged from the wrecks and Lieutenant Eberhardt, the flight commander, obtained consent from Germany for four senior Spanish pilots to fly them. They went into action over the Guadarrama mountains to the north of Madrid at the end of August. The Rambaud Squadron's first action took place on 23 August 1936 when three He-51s escorted Ju-52s sent to attack Getafe aerodrome south of Madrid. They turned for home, where Captain Rambaud and Lieutenant Ramiro Pascual's He 51s were damaged on landing. The fighter had a tendency to bounce on landing and veer once it was on the ground. One of the damaged He-51Bs suffered a broken propeller for which there was no replacement. Later, the resourceful German mechanics repaired the propeller using what was on hand at Tablada, and the aircraft was ready for operations once again.

With a sense of increasing frustration, the German contingent demanded that only they should be allowed to fly the Heinkels. Eventually, the Escuadrilla Rambaud was disbanded. The German pilots requested that they be allowed to engage in combat operations, and this permission was granted.

The Germans go into combat

On 24 August, the He-51s went into combat with German pilots at the controls, the first time that Germans had fired their guns in action since 1918. The heat over the centre of Spain was overpowering. Lieutenant Hannes Trautloft recalled that he wore shorts and a T-shirt. He, his commanding officer Kraft Eberhard and Herwig Knüppel gazed down at the wooded hills and mountains of the Sierra de Guadarrama, where Madrilenians go to escape from the 40+ degrees of the summer city, then a battlefield where, between the trees and crags, Republican militia bussed out from Madrid fought General Mola's conscripts, blue-shirted Falangists from the cities of Old Castile and red beretted Carlist militia from Navarre.

Trautloft shot down a Bréguet XIX bomber, while Eberhardt downed a second Bréguet during the same encounter. The following day saw Knüppel and Eberhardt dispatch one Bréguet each, and on August 27 Knüppel shot down a Nieuport fighter. Two days later Eberhardt destroyed a Potez bomber, one of those which had been supplied by the French government and were flown by mercenaries (see Chapter 4). Finally, that month Ju-52s bombed the War Ministry in central Madrid.

The date, 30 August 1936, brought both good and bad news for the German pilots. Knüppel, Eberhardt and Trautloft each destroyed a Potez 54. As Trautloft fired at his victim and scored a hit, machine gun bullets peppered the right wing of his Heinkel, sending the biplane into a spiralling dive. His controls shot away, the pilot hauled himself out of the cockpit and opened his parachute at about eight thousand feet. The Republican fighter who had shot out his controls returned to finish the job and began firing at the dangling German, but Eberhardt and Knüppel chased the enemy away. Having been the first German fighter pilot to shoot down an enemy plane over Spain, Hannes Trautloft gained the more dubious distinction of being the first to be shot down. He downed a Nieuport fighter on 1 September to even the score.

By 28 September 1936, the Heinkel pilots' scores were as follows: Knüppel six victories, Eberhardt five, Trautloft three, von Houwald three, Hefter one and Klein one, a total of nineteen. Thus, with well-trained pilots, even the rather out-of-date Heinkel 51 was a match for Spanish aircraft. However, on 17 September 1936, a lone Republican Dewoitine 372 fighter, recently supplied from France, accompanied by a single Hawker Fury, two aircraft which were more advanced than the Heinkel fighters, forced them to scatter. The Germans now recognized that they could not always assume that they had air superiority while flying the He-51.

On 28 September, Lieutenant Eberhard Hefter was a member of a low-flying Heinkel team, flying over the Navarrese town of Vitoria, when his machine lost altitude because of engine trouble. One of his wings struck Vitoria's town hall tower, sending the plane crashing to the street, and Hefter became the first fatality of the Heinkel squadron. On 30 September, Kraft Eberhardt and Trautloft downed one Potez bomber each, for their sixth and fourth victories, respectively.

In the meantime, a further shipment arrived on from Germany on the *Wigbert*. The Republican Navy was attempting to blockade the harbour of Cádiz and the river port of Seville, so the *Wigbert* docked at Lisbon and its cargo, by the good offices of Prime Minister António d'Oliveira Salazar, whose regime was friendly to the Spanish Nationalists, forwarded the material to Francoist Spain. This consisted of two Junkers reconnaissance planes, six more He-51 fighters, bomb racks for the Ju-52s, 21,000 tons of bombs and 150,000 rounds of machine-gun ammunition for the fighters. Heinkel fighters, at a strength now of fourteen machines, took on the task of escorting bombers on missions. When Lieutenant Oskar Henrici shot down three Republican machines on 14 October, it demonstrated that a combination of skill, training, first-class work by the mechanics so that engines did not fail and guns did not jam, ensured that the He-51 was still a very useful fighter.

General Wilberg, in charge of the whole process of supplying material to Franco, visited him in mid-August, with the result that further He-46 short-range reconnaissance aircraft were dispatched to Spain.

However, now that the original task of airlifting troops from Morocco was coming to an end and German airmen were undertaking combat missions over Spain itself, a stage had been reached in German intervention in the Spanish Civil War which required considerable reorganization.

Colonel Warlimont and 'Operation Guido'

The number of German aircraft and German combat operations in Spain were such as to be beyond the scope of Major Von Scheele's command. On 31 August 1936, he was replaced by Lieutenant Colonel Walter Warlimont, who would be General Wilberg's direct representative at Franco's headquarters. Specifically, Warlimont was ordered:

1. To examine all possibilities and proposals for supporting the Spanish insurgents, now generally known as Nationalists.
2. To advise the Nationalist High Command.
3. To bear in mind German interests in military, political and economic matters.
4. To cooperate with Italian forces aiding Franco.[26]

Warlimont left Germany on 1 September. He flew to Rome and met his Italian opposite number, Colonel Mario Roatta. Boarding a ship at Gaeta and clad in Italian uniform to conceal his presence, he sailed for Spain. From Seville, he flew in a Ju-52 to Franco's temporary headquarters at Cáceres where the Spanish general was directing the advance of his legionaries and Moors through Extremadura.

Warlimont reported that considerable numbers – an exaggeration – of French aircraft were being dispatched to the Republic and recommended an increase in German supplies. Soon anti-tank units arrived, together with Mark I Panzer tanks under the later general Von Thoma, who would gain distinction in the North African campaign in the Second World War before being taken prisoner.

Major changes, however, were taking place. On 30 September 1936, General Franco became generalissimo, with absolute command over all the Nationalist forces, as well as head of the new government and of the new State. On 3 October he informed Hitler of his status, sending him good wishes and expressing his gratitude. The Fuehrer did not reply because, as the Assistant State Secretary, Hans Dieckhoff, cabled Du Moulin, the German ambassador in Lisbon for communication to Franco, any apparent recognition of the new Spain would 'compromise our work in Spain'. Germany would extend de facto recognition when Franco took Madrid. Du Moulin visited Franco at his new headquarters in Salamanca on 6 October. The generalissimo, reported Du Moulin, said he understood the German position perfectly.[27]

By the end of October 1936, Germany had provided Nationalist Spain with 146 fighters, bombers, reconnaissance planes and He-59 and He-60 seaplanes. The task of the seaplanes was to observe the movements of the Spanish Republican fleet and to track shipments arriving in the Mediterranean ports, which, from somewhat west of Málaga as far as the French frontier, were under control of the Spanish government.

Germany was sending large quantities of bombs and anti-aircraft batteries (including the 88mm high-velocity cannon, which would be famous in the Second World War) to keep the Francoist war effort going. By now there were two flights or *Kette* of bombers, formed from six of the Ju-52 transports, fitted up as bombers, and known respectively as 'Pablos' and 'Pedros', according to the pseudonyms adopted by the German pilots. Some were being flown by Spanish airmen. The German bombers were effective in destroying enemy concentrations, consisting largely of untrained militia, who were steadily falling back on Madrid. They destroyed a fuel depot at Alcázar de San Juan, a railway junction south of the capital where the lines to Andalusia and Valencia diverged. A few bombs dropped from low height (because the enemy had no training in anti-aircraft measures) easily dispersed Spanish militia assembling to try to reduce the fortress or Alcázar of Toledo. The 'Pablos' also helped to relieve the northern city of Oviedo, held by insurgents and besieged by Republican militia.

The 14 He-51 biplane fighters, now called the Gruppe Eberhardt after their commander, were directed from their headquarters at Cáceres, about 185 miles from Madrid, but were operating from Talavera, about 90 miles closer to the capital. The He-51s, some still flown by German pilots while Spanish flyers were being trained, were not effective against anything more modern than the Bréguet XIX with its maximum level speed of just over 100 mph and the Republican Nieuport-32, although the latter with its 125 mph maximum level speed might, with a skilful pilot, be a match for the He-51. In a battle over Talavera, two Nieuports of the Republican Air Force could not manage to shoot down a He-51, but the ace Nationalist pilot Julio Salvador flying the Heinkel could not down either of the Nieuports either. In mid-September 1936, the German Heinkels were flying a mission in direct support of friendly ground forces advancing along the Tagus valley when they were called upon to conduct a low-level strafing mission against Republican infantry. Lieutenant Knüppel recalled:

> We flew daily to-and-fro between Cáceres, Navalmorales and Talavera, and accompanied the Spanish columns in the Tagus valley as they advanced on Madrid. It was here that Trautloft and Houwald brought down some enemy light bombers. This was greeted in especially lively fashion by the brave Moroccans [...]. At Navalmorales forward airfield, the Moros supplied us with tea and mutton when, after our first flight to the front in the Talavera region, we made an interim landing there for breakfast.[28]

Concerned for their Statute of Autonomy, granted on 6 October by the Republic, which Francoist military ideology hated, the three Basque provinces with their capital at Bilbao on Spain's north coast had defeated the military insurgents.

On 25 September 1936, Ju-52s had bombed Bilbao. The German bombers had taken off from Vitoria, some 30 miles south. Vitoria was capital of Navarre, the Carlist or Traditionalist centre which had declared openly for the insurgents. Lieutenant Colonel Warlimont, whose mission was now known under the pseudonym of 'Guido', asked for reinforcements, including twenty-four additional He-51 fighters for the Spaniards and twelve more for the Germans, as well as three experimental Messerschmitt Bf-109 monoplane fighters, which in their later developed models would be a match for British Spitfires. Three experimental Henschel dive-bombers also arrived. Spain was becoming a testing ground for ever more advanced German aircraft. Three more Ju-52 bombers came also. These would join the German bomber squadron together with the Junkers which had arrived in late July and early August 1936 and were now no longer needed for the airlift. Seaplanes also arrived to protect eleven German cargo ships which, between 26 August and 5 October 1936, left the ports of Hamburg and Stettin bringing the large amounts of rifles, machine guns, 88mm anti-tank guns, ammunition, bombs and radio equipment needed, as well as lorries to transport the expeditionary force from airfield to airfield as Franco's forces advanced towards Madrid. A further total of 389 officers and men came down the gangway in the port of Seville, on the broad River Guadalquivir.

So, by the end of October 1936, General Franco's Nationalist insurgents could count on twenty-four Ju-52 bombers, two He-70 medium bombers and thirty He-51 fighters, as well as the small experimental dive-bomber squadron. All came under the authority of Lieutenant Colonel Walter Warlimont.

Nevertheless, by this time it was becoming clear that an even greater German effort would be needed to counter the significant supplies which were reaching the Republican zone from the Soviet Union. Italy had also sent a force of aircraft to help Franco. It was not in Germany's interest to allow the Popular Front regime in Republican Spain to crush the Franco insurrection with Soviet help, nor for Franco to depend overmuch on aid from Italy.

The Condor Legion

The relatively small and not over-modern German force in Spain was being called on more and more to support the limited number of Nationalist aircraft in the battles around Madrid. Fourteen fighters were based at Avila with some reconnaissance-tactical He-46 machines; the six Ju-52 bombers, known as 'Pedros' and 'Pablos', were at Salamanca. Two Junkers, used as couriers, were still flying the Tetuán-Seville-Salamanca route, and there were two seaplanes based at Cádiz and experimental dive-bombers based at Jerez de la Frontera. In all, there were thirty-three German-flown planes. But now aircraft had to be rapidly diverted to raid the Mediterranean ports of Republican Spain where

Soviet supplies were being landed. On the night of 26–27 October 1936, five Ju-52s of the 'Pedro' and 'Pablo' flights, having flown from Salamanca south to Seville and then Granada, bombed the Mediterranean naval base of Cartagena.

Colonel Warlimont flew to Berlin to warn of Soviet arms supplies, among them modern aircraft, which were now arriving to aid the Spanish Republic, though at that time information was vague. Towards the end of October 1936, the German government learned that the Soviet Union was definitely sending war material to aid the Spanish Republic. At meetings of the Non-Intervention Committee in London, Russian diplomats had declared that the USSR could not stand back and see Germany and Italy arming Franco despite having agreed to observe non-intervention. From the German point of view, it was urgent to end the war by taking Madrid. The Non-Intervention Committee could then be wound up and international tension could be reduced once the Germans had got what they wanted, namely a Franco victory.[29]

However, on 28 October 1936 events moved forward dramatically, when Franco and the Germans were taken completely by surprise from the air. Four fast bombers, of a type never seen before, swooped down on Tablada air base near Seville, dropped bombs and sped off at a rate that no Francoist fighter could match. These machines were of the Russian type SB (*Skorostnoi bombarbirovshchik*), or 'fast bomber' designed by A. N. Tupolev in the USSR in the early thirties. A cantilever monoplane with all-metal stressed-skin construction, with a retractable undercarriage, fully flushed rivets and an enclosed cabin, the '*Katiuska*', as it became known in Spain, after a Russian character in a Spanish musical comedy of the time, was very advanced, though some of its complexities had had to be abandoned for the sake of mass production. Its deficiencies would become obvious later in the Spanish war. The Soviet air force was equipped with four hundred of these fast bombers, enjoying a top speed of 265 mph. One Soviet-sized squadron of thirty-one SB-2 *Katiuskas* arrived in Spain in October 1936, with its crews, and was divided into three Spanish squadrons or *escuadrillas*.[30]

The *Katiuskas* raided Nationalist airfields on 30 October and 1 November, destroying Italian fighters parked in the open. The official German record says, with stark brevity, 'With their new type of aircraft, a new phase of the air situation started.'[31] Even the fast and manoeuvrable Fiat CR-32 fighters supplied by Italy might not be able to tackle them.

Furthermore, Soviet Polikarpov I-15 fighters now began to fly missions against the Nationalist columns converging on Madrid from the south-west.

In mid-October 1936, Major Hermann Plocher, a German general staff officer, found himself posted to *Sonderstab W*. His task was to draw up plans for the creation of what was to be known as the Condor Legion. Now, *Unternehmen Feuerzauber*, the transport of Franco's forces to Spain from Morocco, and '*Guido*', the code word for Warlimont's operations, were to be replaced by

Winterübung Rügen, suggesting a winter exercise around the large island of Rügen on Germany's Baltic coast. On 30 October the German War Minister, Field Marshal von Blomberg, and the Foreign Minister, von Neurath, instructed Admiral Canaris, Head of Intelligence, and General Hugo Sperrle, appointed to lead the augmented German force now to go to Spain, to tell Franco and Colonel Warlimont that a decision of great moment had been taken: Germany was to commit itself to a major increase in aid to Franco in his struggle to overthrow the Spanish Republic, in the form of a complete expeditionary force. Germany was willing to supply more aircraft, but their leadership would be in the hands of a German officer, who would answer only to Franco. Franco of course held strategic command, but the German forces would operate as a single force under German tactical command. General Sperrle, the appointed commander, who would later command an air fleet in the Battle of Britain, would report from Spain directly to *Sonderstab W* and to the *Wehrmacht* high command. All pilots, anti-aircraft and rear echelon personnel already in Spain would be incorporated into the German corps to be formed. If Franco accepted these terms, the following reinforcements would go to Spain:

1. A bomber group (*gruppe* should perhaps be translated as 'wing', that is a number of squadrons)
2. A fighter group
3. A long-range and a short-range reconnaissance squadron
4. Two signal companies
5. Two maintenance companies
6. Three heavy anti-aircraft (AA) batteries, often called *Flak*
7. Two searchlight platoons

Personnel, aircraft, anti-aircraft (*Flak*) batteries and all other material to support the 'exercise' were to come from existing Luftwaffe units. Only the very highest quality of men were to be selected. Some were volunteers, while others were 'volunteered'.[32]

According to Warlimont's personal statement to the American historian Raymond Proctor, 'General Franco had not asked for it and indeed did not want it.'[33] Franco may well not have asked for an increase of that size. Perhaps he had been advised that such an augmentation in German aid would only encourage the Soviets to increase theirs. The German high command was also anxious to limit German support for Franco to a level which would not provoke French intervention on the side of the Republic. Specifically, German pilots were ordered to keep 30 kilometres away from the French frontier.[34] One might speculate that Franco feared that the price to be paid in concessions to Germany might prove too high or that a victory over the 'Reds' with Nazi aid would have a divisive effect on Franco's conglomerate of different ideological forces, united only in their determination to overthrow the Popular Front government of the Spanish Republic. The somewhat arrogant tones of the German message could hardly have failed to irritate the Spaniards, for the instruction to Canaris read: 'The following point of view is to be expressed most emphatically to General Franco', while its third paragraph ended the instruction, 'On the condition that General Franco recognises these demands without reservations, a further activation of German aid is envisaged.' The 'demands' included that the Spanish general should change his 'hesitant and routine procedure' (which included the 'scattered employment of his aircraft'), and that Franco should employ a 'more systematic and active conduct of the war with regard to ground and aerial operations'.[35] Nevertheless, in the event, the Condor Legion commanders maintained a very tactful relationship with the Spanish high command, never trying to interfere with strategic decisions but carrying out Spanish tactical orders faithfully. That the Condor Legion's future chief of staff von Richthofen soon learned Spanish is an indication of how well the German authorities in general understood how to get on with the prickly pride of Spanish officers. Berlin would recall General Faupel, the Nazi ambassador to Franco Spain, because he was angering Franco and his entourage by seeking to give unrequested advice on political matters.[36] Similarly, the Soviet Union would be widely resented among the Spanish Republicans for its attempts to influence internal affairs.

The new German force was code-named *Eiserne Rationen* ('Iron Rations'), then *Eiserne Legion*, before Goering, the Air Minister, insisted on *Condor Legion*. Apparently, Goering wanted all Luftwaffe units designated by bird names and, while this proved impractical, the German corps in Spain retained its avian title. The units of the Condor Legion were designated by abbreviations of their function, followed by a slash and the Nazi symbol 88 indicating Heil Hitler! (H is the eighth letter of the alphabet). Thus, the bomber group (*Kampf*) would be known as K/88, the fighter group (*Jagd*) would be J/88, the reconnaissance group (*Aufklärungs*) A/88, anti-aircraft (*Flak*) F/88, the seaplanes (*Aufklärungssee*) *AS*/88 and the signals (*Luftnachrichtenabteilung*) LN/88. Staff was to be *S (Stab)*/88. The Condor Legion's commander was to be *Generalmajor* (equivalent to British brigadier or US brigadier general) Hugo Sperrle, with Lieutenant Colonel Holle as chief of staff.

Sperrle was a huge, physically dominating man, with a brutal countenance, who had been in command of all the air forces of the German 7th Army in the First World War and had rejoined the Luftwaffe in 1935. He was considered a highly competent commander. Within two months, Lieutenant Colonel Baron von Richthofen replaced Holle as chief of staff. A successful fighter pilot in the First World War, he was a graduate engineer who had been in charge of procurement and development for the new Luftwaffe. Able to make snap decisions, surviving Condor legionnaires told an American historian that von Richthofen was liable to express his frustrations with the more dilatory Spaniards.[37]

The early order of battle of the Condor Legion was three squadrons of twelve Ju-52 bombers, to which would be added the Junkers bombers in Spain already as well as the experimental flight of dive-bombers. F/88, the fighter unit, was composed of three squadrons of nine fighters plus a fourth squadron formed with the fighters already in Spain. Reconnaissance (A/88) had He-45 and He-46 machines for short-range work and He-70 and Dornier Do-17 for long-range scouting, with a total of twelve aircraft. The seaplane flight already in Spain became a reconnaissance-bomber squadron with fourteen aircraft. Counting the material already in Spain, the Legion included about 119 aircraft.[38] There were twelve anti-aircraft (F/88) batteries, including five batteries with four 88mm guns each, and thirty lighter guns of 20 and 37 mm, a mixed-calibre battery for training Spanish gunners and an ammunition column. Signals was composed of five telephone, radio and specialist aircraft location companies. The force had maintenance sections for aircraft and land vehicles, photography, medical services, field police and sundry others, thus creating a self-contained air unit. By the end of the Spanish Civil War, the Condor Legion would have a strength of 5,136 officers, men and a few civilians.

Sperrle left Berlin for Spain in a Ju-52/3m on 31 October, travelling via Rome, while the new personnel of the Condor Legion, still under the impression that they were to carry out a training exercise in the Baltic, left Stettin in winter uniforms. When the ships turned west into the setting sun they began to wonder where they were going. As they entered the English Channel, they were told to change their uniforms for civilian clothing. In that busy marine thoroughfare who knows what inquisitive ship or low-flying aircraft would make of men in uniform on deck? Lieutenant Douglas Pitcairn, who despite his name was a German Luftwaffe pilot, recalled that they were told that they were on their way to fight in Spain only when they approached the Bay of Biscay.[39] They docked at Cádiz and were entrained for Seville, where they were issued with Spanish uniforms.

The K/88 bomber crews were to fly their machines to Spain. They were assembled at Lechfeld near Augsburg, flew to Ciampino near Rome, thence to Cagliari on Sardinia, to Majorca and finally to Seville, whence they were sent on to Salamanca and León. They were now to meet the challenge of the advanced planes and well-trained pilots of the Soviet air force.

Chapter 3
Fascist Italy's aircraft create an international crisis

Spanish army plotters against the Republic had sought Italian Fascist aid before the civil war. In April 1932, the monarchist officer Major Juan Antonio Ansaldo had asked Marshal of the Italian air force Italo Balbo to support General Sanjurjo's attempt at a coup d'état. Two hundred machine guns were promised but the coup was a fiasco and the weapons never arrived. A year later, Ansaldo and the right-wing leader José Calvo Sotelo met Balbo again to seek support for a coup. Again, weapons were promised but none arrived. On 31 March 1934, Antonio Goicoechea, leader of the right-wing *Renovación Española* party, and a group of monarchists, together with General Emilio Barrera, once more approached Mussolini, the Italian Fascist dictator, and solicited funds, arms and training in Italy.[1]

On 1 July 1936, Pedro Sainz Rodríguez, a monarchist conspirator and later Minister of Education in Franco's government, signed four contracts in Rome for sizeable and very detailed quantities of arms, including Savoia-Marchetti bombers, Fiat CR-32 fighters and Macchi seaplanes.[2] Despite these very recent contracts, Mussolini delayed for a week after the beginning of the military uprising while the journalist Luis Bolín, who had been sent to Rome by Franco from Morocco to negotiate for aircraft, talked with the Italian Foreign Minister, Mussolini's son-in-law Count Ciano. Mussolini, however, rejected Franco's appeal for arms. Now that the uproar aroused by his recent aggression against Abyssinia seemed to have quietened down, this was not perhaps the moment to awaken suspicions in Britain. There was also the question of how the new French Popular Front government might react to any Italian involvement in Spain.[3] However, plans for expansion of the Italian air force, known as the Regia Aeronautica, looked forward to an air force of over three thousand machines. Thus, the air force resented being treated as a mere auxiliary service to the army. Spain offered a fine opportunity to spread its wings.[4] In Italy, the spirit of idealism and the association of flying with honour, duty and ultra-nationalism was strongly encouraged by the Fascist outlook and was linked with the Italian preoccupation with speed associated with the ultra-nationalistic movement founded in 1909, known as Futurismo, which admired speed, technology, violence and the new machine of speed, the aeroplane. Even more than the Luftwaffe, the Regia Aeronautica would respond favourably to Franco's request for aid.

Franco had used another channel when he sought aircraft in Germany and came up against the refusal of the Wilhelmstrasse. Now, during that first week of the Spanish Civil War, the Italian military attaché in the internationally run city of Tangiers, Major Giuseppe Luccardi, whom Franco, only 37 miles away in Tetuán, persuaded to his cause, cabled Rome on his behalf several times.[5] At first Mussolini refused to attend to Luccardi's recommendations, among other reasons because reports were reaching Rome that the French Popular Front government was considering sending aircraft and other equipment to Spain. A clash with France was not desirable. Ciano continued the exchange of cables with Luccardi, coming round to the view that he should persuade Mussolini to agree to aid the Spanish insurgents. A delegation under Antonio Goicoechea also waited on the Duce, but Mussolini hesitated because he did not yet know for certain that Franco and the Navarrese Carlist militias who formed the bulk of General Mola's militia in Pamplona were allies. Nor was he sure that Franco was actually in control in Morocco. The Italian dictator decided to accede to Franco's request only when it became clear that centre and right-wing opposition in France to aiding the Republic was so intense that not much French help would actually reach Madrid and Barcelona. Always under the effect of the inferiority complex which Italy suffered in the face of British and French imperialism, together with his rivalry with Nazism, Mussolini calculated that the Spanish war, hopefully, would be brief, that there was obvious practical advantage in testing the airmen and the aircraft and that a hostile power in Spain would clip French wings in the Western Mediterranean. From the evening of Friday, 24 July, as Franco's emissaries were landing in Berlin on the eve of their momentous meeting with Hitler, the Italian Ministries of War and Aviation began to organize the dispatch of material to Franco's forces in Morocco. On 27 July, the freighter *Emilio Morandi* sailed from the naval base of La Spezia bound for Melilla on the northern coast of the Spanish zone of Morocco. It was loaded with bombs, fuel, ammunition and spare parts. Germany's first ship would leave Hamburg for insurgent Spain a few days later, though some Junkers-52 planes had flown direct to Seville. The first items of German and Italian aid were thus dispatched to Franco at almost the same time.

Hitler took his own decision regarding help for Franco. Similarly, neither the King of Italy, nor General Badoglio, the chief of staff, nor General Valle, Undersecretary for Air and chief of air staff, had any part in the decision to intervene in Spain or to coordinate or control the Italian forces involved. Decisions of this nature were handled by Mussolini, as head of the government and Minister of War and Air, by Galeazzo Ciano as Foreign Minister and by the *Ufficcio Spagna* set up within the Foreign Ministry, which received all communications from the Spanish campaigns. In Spain, the commanders of the *Aviazione Legionaria* were subordinated to General Mario Roatta, an army general who saw the air force as a substitute for artillery and tanks rather than as an independent strategic weapon, while the *Aviazione delle Baleari* – the Italian air force on Majorca – was directly under the command of Mussolini and Ciano.[6]

On Tuesday, 28 July, 38-year-old Lieutenant Colonel Ruggero Bonomi, who had experience of long-range flying, arrived at the Italian Air Ministry, having been summoned by General Giuseppe Valle. The latter offered him a special mission in Spanish Morocco, which Bonomi accepted at once. In command of a squadron of twelve Savoia-Marchettti S- 81 bombers, a type nicknamed *pipistrello* (bat), which were to be assembled at Elmas aerodrome near Cagliari in the south of the island of Sardinia, he was to fly to Melilla in Spanish Morocco and put himself, the crews and the aircraft at the disposal of General Franco. These up-to-date three-engine aircraft, with five crew members, had a cruising speed of 225 mph, which was unrivalled in the Spanish war until the arrival of the Russian SB-2 *Katiuska*, and could carry a load over two tons of bombs in various combinations ranging from small incendiaries to 250-kilogram high-explosive missiles. They were equipped with three pairs of machine guns in the front, others firing downwards and from the sides. The flight, of about 750 miles, was to be completed in four hours thirty minutes without stops and would be feasible for aircraft such as the S-81, which had a theoretical range of up to 1,250 miles.

Bonomi selected Major Altomare and Lieutenants Lo Forte and Erasi, all first-class navigators, to lead three of the four flights of three aircraft each, while Bonomi himself would lead the first.[7] Early on the morning of Wednesday, 29 July, they flew to Cagliari. At 10.45 am, the twelve Savoia-Marchetti bombers roared in from their bases. Bonomi informed the crews of their task. They would, if necessary, stay in Morocco to instruct Spanish airmen about the characteristics of the *Pipistrello*. While there, they would enrol in the Spanish Legion to avoid any international complications.

That Wednesday evening, 29 July, a seaplane splashed down at Elmas and General Valle and the Spanish journalist Luis Bolín emerged accompanied with Ettore Muti, a Fascist leader with a distinguished record in the First World War. Muti was there as personal emissary of Count Galeazzo Ciano, Mussolini's son-in-law and Italian Minister of Foreign Affairs. General Valle dined with the crews who were to fly the Savoias to Franco and told Colonel Bonomi that Mussolini had given the green light to the expedition. The twelve aircraft were to leave the next day. They were to fly due south at a height of six thousand feet and when they hit the Algerian coast they were to fly west at a greater height as far as the Chafarinas, three islets in the Alborán Sea off the coast of Morocco, and then straight to Nador, the airfield of Melilla, just inside the Spanish zone of the Moroccan Protectorate. Two ships had been stationed along the route to help them, and Valle would fly ahead to verify meteorological conditions. These precautions, nevertheless, did not prevent serious accidents and loss of life of some of the crews of the twelve Italian aircraft.

Finally, at 5.35 on the morning of Thursday, 30 July 1936, the twelve Savoia-Marchetti S-81 bombers, drawn from squadrons of the Royal Italian Air Force, took off from Elmas aerodrome, heavily loaded with fuel, and climbed slowly into the dawn sky.[8] Unlike the silent arrival of Goering's Junkers transports, these Italian aircraft would become front-page news and lead directly to Britain and the rest of Europe formulating the notorious policy of Non-Intervention in the Spanish Civil War. Italy would go on to provide six thousand air personnel, among them 1,435 pilots, and dispatch 764 aircraft, apart from the over seventy thousand infantry, artillery and other ground troops who would form the *Corpo di Truppe Voluntarie* or CTV, which fought on Franco's side.[9]

Luis Bolín, the Spanish monarchist journalist domiciled in London, sat in the rear gunner's turret as the twelve Savoia-Marchettis, in four flights of three machines each, cruised in V formation and line astern high across the Western Mediterranean towards North Africa. He could be well-satisfied, for he had arranged Franco's safe flight from Las Palmas to Tetuán and now he had helped persuade Mussolini also to come in on the Spanish general's side.

The technical aspects of the flight from Sardinia to Morocco were, however, problematic. The Italian aircraft encountered fierce headwinds, together with cloud that separated the flights from each other. Speeds fell worryingly. Low cloud prevented the navigators seeing the ships which had been stationed to help them fix their positions. After five hours' flying time, the Savoias were off Oran, far behind their planned position. Fuel tanks were getting low. At 10.50 am, Lieutenant Angelini's aircraft fell away from Lieutenant Altomare's flight and did not respond to a radio call. It was last seen flying south but was soon lost to view among the clouds. It crashed into the sea about 50 miles north of Oran. At 11.29, its wreckage was sighted by an aircraft of the Belgian Sabena line. It was floating and the Belgian crew saw three survivors clinging to a wing. The Sabena plane broadcast an SOS but rescue ships, among them the French patrol ship *Giselle*, sent urgently out of Oran, could not reach the downed aircraft in time to save the crew. At 11.45, Colonel Bonomi's aircraft landed at Nador, followed by eight other S-81s. The flight had taken six hours and ten minutes, one hour and forty minutes longer than estimated. The two aircraft flown by Lieutenant Mattalia and Captain Ferrari did not arrive. The former had run out of fuel and had tried to land at Saidia just inside the French zone of Morocco. Only one of his crew survived. Ferrari, however, had made a skilful emergency landing, though two crew members had been killed, on the French side close to the outlet of the Muluya river, which formed the frontier between the French and Spanish zones of Morocco.

As soon as Bonomi and the eight Savoias landed at Nador, a Spanish Bréguet XIX flew off to search for the three missing aircraft. The search was unsuccessful, so Bonomi ordered Lieutenant Baduel to take off, given that his fuel tanks still contained a few gallons. He saw the two downed aircraft, whose position in the French zone had been notified by radio by General Valle, who was now on his way back to Sardinia.

No instructions about what to do or say in case of an emergency or crash landing had apparently been given to the surviving Italian crew, who were interrogated by the French authorities. Although the aircraft had been repainted in grey, the Italian insignia had been only lightly painted over and it soon became evident that the aircraft were Italian warplanes destined for Franco. This was despite the fact that the crews had been given false names and were wearing civilian flying clothes. One, presumably disobeying orders, had retained a document identifying his squadron in the Regia Aeronautica. The French searched inside the Savoia and found six machine guns as well as Italian military greatcoats. Without realizing that the French authorities were on the scene, two Spanish insurgent aeroplanes flew over the area and dropped parcels with Spanish Legion uniforms, with a written order to the supposed legionnaires to tell the French that they had lost their way while carrying out a reconnaissance of the territory. The information, confirmed in a dispatch from the French Resident-General in Rabat,[10] created a scandal for the readers of the French press as they were about to leave for *les grandes vacances*, for, only a month earlier, international sanctions imposed on Italy over its aggression in Abyssinia had been lifted. Now Mussolini appeared to be backing an insurrection against a European member of the League of Nations. Was another crisis building up?

In France the right wing blamed Léon Blum, the Prime Minister, for agreeing to arm the Spanish 'Reds' and so provoking Mussolini. In contrast, Blum would later recall that the news that Italy was arming the Spanish insurgents made him and his colleagues less nervous about helping the Republic.[11] However, pressure from the British government, from much of the French press and from many of his own colleagues, both socialists and radicals, who warned him that France could not afford to lose British support if indeed foreign intervention led to a European war, led him to propose, with the full support of the British Foreign Secretary, Anthony Eden, a policy which would become known as Non-Intervention. It was not a treaty. The French Foreign Ministry merely asked every European country to forbid its nationals to export war material to either side in Spain. All countries (save neutral Switzerland) agreed. Although Germany and Italy could not refuse the French proposal, they had no intention of abandoning Franco, so they continued to supply the Spanish general with war material, particularly aircraft, which were of inestimable value to him, especially as their numbers were steadily increased, while, as will be seen, a small and finite number of aircraft of doubtful value and without logistic back-up were flown to Spain from France. Not till October 1936, when the Non-Intervention agreement was clearly seen to be working against the interests of the Republic, did the Soviet Union, whose Popular Front policy dictated that it should not be seen to be supporting the revolutionary situation in Republican Spain, at last decide that it could not abandon the Republic and that it would send bomber and fighter squadrons, which effectively saved Madrid.[12]

Nine of the twelve Italian planes, consequently, landed just inside the frontier of the Spanish zone of the Moroccan protectorate, which was completely under Franco's control. Colonel Bonomi ordered the fuel which remained in eight of them to be syphoned out and put into one machine in which he and Bolín flew west to Tetuán, while the other Savoias waited for a tanker to arrive at Melilla with the high-octane fuel that they needed and which could not be found in Spanish Morocco, a fact which had inexplicably not been considered earlier. Bonomi, presumably obeying instructions from the highest authority, ordered all his men to enrol in the Legion, also known in Spanish as the *Tercio de Extranjeros* or 'Regiment for Foreigners'. In this way, the Italian S-81s were disguised – though nobody was fooled – as the *Aviación del Tercio*, which had not had an air arm previously. At 9 am on 3 August, the *Emilio Morandi* dropped anchor in Melilla harbour with the high-octane fuel required. Next morning the Savoias flew to Sania Ramel airfield in Tetuán, which was to be the Italian air base for operations in the Strait of Gibraltar, though one Savoia's engine malfunctioned and crashed into a Bréguet XIX on the ground, an accident caused by the cloud of dust thrown up by the propellers. At Tetuán they were inspected by Franco's air force commander, General Kindelán, while the band of the Legion played suitable martial music. Franco gave a brief harangue to the Italians drawn up by their aircraft. Then he shook each man by the hand. This made a great impression on Colonel Bonomi, who wrote in his report to Rome that 'the arrival of our mission has created an impression in the Spanish milieu not less than the undersigned noticed when the admiral commanding the German ships at Tangiers visited General Franco'.[13] The German was Admiral Rolf Carls, who flew his flag in the pocket battleship *Deutschland*. While ostensibly German ships were off Spanish coasts to protect German citizens, their presence was intended to inhibit and to spy on the Republican fleet's movements.[14] Bonomi's comment displays to what degree the Italians were concerned to rival the German contribution to Franco's uprising, for reasons as much apparently of prestige as of political advantage.

The loss of the three Savoias en route from Sardinia to Melilla on 30 July may well have embarrassed the Italians, considering that so many German Ju-52s had flown safely much further. Bonomi soon had the chance to demonstrate the ability of his planes and crews when on 8 August two Savoias were sent hastily to Larache, on the Atlantic coast of Morocco, to put to flight the Republican destroyer *Almirante Valdés*, which was shelling the port and town. However, despite all their efforts, the Republican destroyer avoided the bombs which the two Savoias tried to drop on it.

During 4 August, a caravan of lorries began to arrive at Sania Ramel airfield near Tetuán with bombs, fuel, ammunition for machine guns and spare parts, which had been hastily unloaded from the *Emilio Morandi*. Like the Germans a week before, in the summer heat the Italians and the Spanish specialist workmen began feverishly to arm the bombs, load machine gun belts and service the Savoias, while Colonel Bonomi discussed details of the Italian role with Franco and his staff in the relative cool of the High Commissioner's palace in Tetuán. Franco's first convoy across the Strait of Gibraltar would require protection from the air, given that the insurgents had hardly any navy while the Republican fleet, with its modern destroyers, represented a grave danger. Two ferries belonging to the line which provided the link between Ceuta and Algeciras, together with two small warships, were to transport three thousand men, six artillery batteries and other military material. For their protection, Franco had four Dornier Wal seaplanes (the others were awaiting new engines), three Fokker VIIs (originally based at Cape Yubi in Spanish Sahara), two Nieuport 52 fighters from Tablada (Seville) and Armilla airfield near Granada, six Bréguet XIX reconnaissance bombers, the DC2 captured in Tablada and two Savoia seaplanes from Cádiz. Together with these, the newly arrived Savoia-Marchetti S-81s, three of which were ready for action, provided a double curtain of protection for the convoy of troops and armament which sailed to Algeciras, just across the bay from the British colony of Gibraltar. Three Savoia-Marchetti S-81s and a further three Bréguet XIXs were ordered to be ready for emergencies.[15] The seaplanes were to skim the water to guard against the submarines in the Republican fleet, the Bréguets and the Nieuports would fly at a maximum height of 1,500 feet, and the Savoia S-81s and the Fokker F.VIIs at a maximum of 4,500 feet. They were to ensure that no Republican warship which was in harbour at Tangiers should leave and none of the Republic's ships at Málaga, including a battleship and two cruisers, should be allowed to sail west of Estepona. In overall air command under Franco, who was to watch the convoy from Mount Hacho, a 669-foot peak overlooking the Strait, was retired general and pilot Alfredo Kindelán.

The planning of the operation was an example of the military principle of concentration of effort for an urgent task and of a careful balancing of risk. On the one hand, the ability of aircraft such as those which the insurgents had at their disposal to launch successful attacks on moving warships was as yet unknown, but on the other hand the Republican ships were known to be crewed by men who had mutinied. Without officers, their morale was thought to be low and their skill at manoeuvre for torpedoing and anti-aircraft defence doubtful.

Early on 5 August, two S-81s flew a reconnaissance over Tangiers to check that the Republican warships were still there. By 6.20 am, all the aircraft were in the air. Towards 7 am they attacked, bombing and machine-gunning two Republican destroyers in the Strait. The *Lepanto* fled into Gibraltar to disembark casualties caused by a 50-kilogram bomb. The authorities on the Rock refused permission to bury a dead sailor, fearing that the interment would encourage a political demonstration among the many refugees on the Rock. Instead of a funeral at sea, the Republican naval command, with little authority over the rebellious sailors, had to consent to the corpse, already decaying in the August heat, being taken to Málaga for burial, when nothing should have interrupted the Republican blockade of the Strait. When the *Lepanto* was escorted out of Gibraltarian waters by British ships, the highly skilled and experienced Spanish aviator and squadron leader, Captain Carlos Haya, bombed it from the same DC-2 which Captain Vara del Rey had immobilized on Tablada air base very early on 18 July when it was getting ready to bomb Tetuán.

When no further danger from Republican ships seemed likely, three rockets with white smoke signalled the convoy to set sail on its 28-mile journey from Ceuta to Algeciras. It was forced to travel at the speed of the slowest ship. After nearly two hours, the convoy was close to its destination when it was intercepted by the Republican destroyer *Alcalá Galiano*. Bonomi, who was with Haya in the DC-2, called on the S-81s to launch all their heavy bombs on the *Alcalá Galiano*. This destroyer, its shell elevator out of action, its anti-aircraft ammunition exhausted and with eighteen killed and many wounded, took shelter in Málaga.[16] The destroyer *Almirante Valdés* was also hit by a bomb from an S-81. Whatever the uncertainty about the effects of bombing moving ships from the air, the somewhat imprecise but intensive bombing from the insurgent aircraft hit not only two Republican destroyers but also the British-registered freighter *Medon* and the Dutch *Zoundewijk*, while five bombs fell close to the Italian cable layer *Città di Milano*.[17] No warning of military activity had been given to merchant shipping in the very busy Strait of Gibraltar.

British Gibraltar enjoyed a grandstand view of the earliest operations of the Spanish Civil War. It had taken in refugees both from the immediate revolutionary outrages caused by news of the military insurrection and from the savage repression launched by Moorish troops and legionaries just over the frontier in La Línea and San Roque and across the bay in Algeciras. On 19 July, some ships of the Republican navy had anchored in Gibraltar harbour in an attempt to buy fuel but had been required to leave by the unfriendly naval authorities of the colony, aware of the mutinies that had taken place aboard them. Observers on the Rock would have seen Franco's few obsolescent aircraft flying handfuls of legionaries and Moors to Seville, while Royal Navy officers carefully watched the events of 5 August. A naval officer had allowed the destroyer *Lepanto* to evacuate its wounded after it had been hit by an Italian bomb but ordered its immediate departure. When the destroyer *Alcalá Galiano* shelled Franco's convoy, it had missed and the shells fell on a cricket ground, interrupting a peaceful game, as *The Times* reported.[18]

This first convoy of Franco's army to sail from Morocco to the Peninsula arrived safely in Algeciras. The officer-less crews of the Republican ships were severely demoralized and panicked by the concentrated attacks, for which they were untrained, which Franco's aircraft had launched over twelve hours. In consequence, the Republican military commander at Málaga, the nearest large port and temporary naval base, insisted that every possible aircraft should be sent to the Strait of Gibraltar to back up the fleet, if it was to prevent Franco's troops crossing from Morocco to the Peninsula.[19] Had this been done in a determined and concentrated manner, the Spanish Civil War might have taken a different course.

The staff work for the crossing of the Strait was highly detailed. Every ship and every aircraft had its own orders. Franco, who observed the operation from Mount Hacho, was kept closely informed at all times. While Colonel Bonomi, whose reports seem to give him more importance than was merited, wanted to risk an attack by the Republican destroyer *Lepanto* when it was required to leave Gibraltar after disembarking its wounded men, the hypercautious Franco delayed the departure of the insurgent convoy with its three thousand Moors and legionaries several times until he was sure as he could be that it would reach the Peninsula unharmed.[20]

Putting into the air almost every aircraft that the insurgents possessed, and the different flying heights and roles identified for them, was an example of the military principle of concentration of effort, while the ragged performance of the Republican destroyers, the absence of its cruisers and of any of its aircraft is an indication of the lack of understanding in Madrid of the importance of blockading the Strait and a reflection of the chaos caused by the military uprising itself. In Madrid there was no real centralization of command, nor had the Republican fleet been reorganized after most of its middle and senior officers had been arrested or killed. The lack of training and experience in avoiding and countering attack from the air meant that the Republican warships were less of a risk to Franco's forces than had been feared, though if more of them had come out of harbour the result of the day might have been different. On the other hand, the insurgents had great advantages. Their aircraft were within minutes' flying time of the Strait. There was no Republican naval base nearer than Málaga, which was at least two hours fast steaming from the Strait of Gibraltar. In addition, the insurgents had coastal artillery on both sides of the Strait, at Ceuta and Punta Carnero, and this also shelled the Republican destroyers.

Successive columns of Franco's expeditionary force were now ready to move north from Seville. Within a few days, the leading columns, leaving a trail of killings in their wake[21], were already at Almendralejo, 105 miles from Seville, and only 19 miles from Mérida, where they were to be joined by General Mola's forces moving south from Cáceres. Only then did the Madrid government decide to send aircraft to Extremadura, the province of central and south-west Spain. On 11 August, Francoist columns commanded by Major Castejón and Lieutenant Colonel Asensio, a later minister in one of Francos' governments, who were both hardened colonial or *africanista* officers, occupied Mérida, aided by three

S-81s which made low-level runs over defensive positions along the banks of the Guadiana river, and overwhelmed the untrained Republican militia with 50-kilogram bombs. Next, the Italians launched three green rockets to signal the waiting Moroccan troops to advance. The militia fled or surrendered, to be bayoneted or shot against the nearest wall after a brief 'trial' for the crime of 'military rebellion'.

Leaving a garrison in Mérida, Colonel Yagüe, the field commander of the expeditionary force, moved west to take Badajoz, a city stormed by the future Duke of Wellington in 1812 during the Peninsular War against Napoleon, and which would now be bloodily taken by the Legion, after which hundreds of Republican militia would be massacred.[22] The attack began with bombing from Captain Carlos Haya's DC-2, followed by those Ju-52s which had been taken off transport duties and were now being flown by Spanish insurgent pilots, one of whom, Captain Francisco Diaz Trechuelo, was killed by a fortunate rifle shot as he flew low over the militias. On 12 August, four S-81s contributed their part by dropping incendiaries and 250-kilogram bombs on the several thousand militiamen lining the ancient walls of the city.

The killing of large numbers of resisters shocked the Fascist Ettore Muti, who wrote personally to Count Ciano,

> Just imagine. The Spaniards are shooting at least thirty people per day in Seville [...] they are judged by a court every one of whose members has had a brother, a father or a son murdered by communists. [...] At Badajoz, all prisoners were shot after a fierce and brave defence.

Importantly, Muti wrote,

> Our Fascism is a much more beautiful thing. (*'Il nostro fascismo è tutto un' altra bellissima cosa'*.)[23]

Nevertheless, it would make no ultimate difference that Franco's uprising was not 'socialist' in the Fascist or Nazi sense but fundamentally a defence of the reactionary Spanish Church, the large landowners and the army. Mussolini, like Hitler, had good reason to wish to see Franco in power in Spain. As Ciano wrote towards the end of the Spanish war, after Italian blood and treasure had been poured out,

> Those silly people who tried so hard to criticise our intervention in Spain will one day perhaps understand that on the Ebro, at Barcelona and at Málaga the foundations of the Roman Mediterranean empire were laid.[24]

Unlike the Germans, the Italian aircraft had received no orders to avoid combat. With Spanish airmen who were undergoing conversion training sitting with them in the Savoias, they carried out the first direct attack on the Republic when on 6 August they bombed a concentration of militia at the railway station at Guadix (Granada) and at the local airfield to which the government had at last sent aircraft, and spent the next few days harassing militia retreating before Franco's Moors and legionaries. They had little opposition from Republican aircraft, most of which were heavily occupied in the Sierra to the north of Madrid. Andalusia and Extremadura were the scene of no more than sporadic individual combats of First World War style between Bréguets and Nieuports, when all Republican effort ought to have been concentrated on making the ports of Morocco and southern Spain untenable for Franco's forces.

Reinforcements

On 14 August 1936, the freighter *Nereide* docked at Melilla carrying a squadron of twelve Fiat CR-32 fighters, with fuel, spares, ground crews and flying personnel, under Captain Dequal. Officially, all were volunteers, which was probably true, though few airmen of the time would have rejected an offer to try out their skills in battle, and particularly not with the slim, silver-painted CR-32. This aircraft, one of the legendary fighters of the 1930s, together with the Messerschmitt Bf-109, bore the brunt of air combat for the insurgents during the Spanish Civil War. The Regia Aeronautica's principal fighter, it was a development of the machine designed by Celestino Rosatelli (thus the 'CR') for maximum strength and manoeuvrability. Dequal's squadron was the first of a total of 377 CR-32s, called '*Chirris*', from the Italian pronunciation of 'CR' as imitated by the Spaniards, to be delivered to support Franco's war effort. It could reach about 215 mph, though later versions were said to have achieved up to 250 mph. Its climbing and diving and recovering ability was outstanding. The insurgent or nationalist pilot José Larios, comparing it with the German He-51, recalled that the Italian plane was heavier and needed firmer handling on take-off and landing but was more robust and faster as well as more manoeuvrable.[25] The Bréguet bombers and Nieuport fighters of the Spanish air force had no chance if they met a *Chirri*, some versions of which had 12.7mm machine guns firing a mixture of explosive, incendiary and soft-nosed bullets, though reports state that not all of these arrived and that sufficient armourers were not sent at the beginning.[26]

The first squadron was flown to Tablada base near Seville within a few days of arrival, assembly and testing in Morocco. Commanded by Vincenzo Dequal, it became the first fighter squadron of the *Aviación del Tercio*, still masquerading as a Spanish unit. Over the 18–20 August, Franco ordered the concentration of all his aircraft, as well as the Italian S-81s and the CR-32s, to disperse Republican columns which were threatening Córdoba. Hundreds of men were bombed and machine-gunned along the main roads.

However, the Republic had acquired some French Dewoitine D.372 fighters and Potez bombers. On 31 August, two out of a patrol of three *Chirris* were downed by two Dewoitines and one of the three British Hawker Furys that Spain had bought before the war. Several more of the *Chirris* were damaged in accidents and landings. When one *Chirri* landed at Córdoba, the dust raised by its propellers led to the pilot crashing into a Bréguet XIX for the second time in a few weeks, while another damaged its landing gear in a pothole.[27] Given the difficult conditions in Spain, more planes were lost by accident than in combat. The CR-32 pilots struggled at first to have an impact on their Republican counterparts because, following a supply oversight in Italy, only two of the dozen Fiat fighters boasted compasses! Unfamiliarity with Spanish terrain and inadequately detailed maps further compounded the unit's navigational problems when in the air, and the end result was pilot disorientation culminating in emergency landings and damaged aircraft. Consequently, Italy sent more *Chirris*. At the end of August 1936, nine were unloaded at Vigo in north-west Spain, for fear of interference with shipping in the Strait of Gibraltar by the Republican navy, and sent by rail to Seville. A further nine followed on 3 September. Three went to Majorca on 27 August and another twelve to Vigo in September. Between thirty-six and forty-five Fiat CR-32 *Chirris* had reached the insurgents by the end of October.[28]

Whatever the high quality of the *Chirri* Bonomi complained about shortages of skilled personnel. Lack of thorough planning and readiness for the eventualities of combat in Spanish conditions was evident. The insufficiencies of the Italian force in comparison with the thoroughness of organization of the German contribution to Franco were striking, and Ettore Muti wrote to Count Ciano about it at length. His view of Italian deficiencies compared with German efficiency is revealing:

> The Germans have marvellous organisation in Spain. They have a large number of ground crew experts and lots of spare equipment. The pilots are all perfect navigators and carry out their duties, night or day, precisely. Their main task is transporting troops from Tetuán to Seville and Jerez. The Spaniards are very enthusiastic about both the Italians and the Germans, but the Germans are the favourites because their machines are always ready. Their aircraft may well be less fast than ours, but they are easier to maintain.[29]

Spain had other surprises for the Italian flyers. On 22 August, three *Chirris* were due at the airfield of San Fernando, near Salamanca in western Spain, from where they were due the next day to escort eight Ju-52s to bomb Getafe aerodrome, south of Madrid, where most of the Republic's fighters were based. San Fernando was a temporary aerodrome or rather airstrip and was not easy to locate. One of the DC-2s, at the time at Cáceres, was supposed to take off in order to guide the Italian fighters. Unfortunately, it did not appear, or perhaps it could not find the CR-2s, which tried to locate San Fernando on their own.

An unexpected storm complicated the situation and the fighters were unable to land. Running out of fuel, they touched down on what seemed from the air to be a suitable field. What Lieutenant Monico and the two other pilots did not know was that the field was used to pasture fighting bulls belonging to one of the famous bull ranches or *ganaderías* of Salamanca. As Monico and the sergeant pilots inspected the damage to their aircraft, the bulls took no notice, continuing to graze peacefully. Suddenly, the Italians heard shouting. They looked up to see a man gesturing and shouting from behind a fence 'Son toros bravos. Vengan!' ('They're fighting bulls. Come this way!'). It took some time for the Italians to realize what he was saying but, fortunately for them, they did so in the end. Once the bulls were removed from the field, the damaged aircraft were loaded on lorries and taken off for a repair which was probably much easier than it would have been had the machines, robust as the *Chirris* were, been charged by angry animals.[30]

That some Italian pilots lost their way was not surprising. Three CR-32s were sent to Granada without any opportunity for the pilots to orientate themselves first. Nor were they given suitable maps of the topography. Sub-lieutenant Cecherelli could not find the airfield at Armilla and had to return and make another attempt the next day. In the end they bought Michelin road maps in a Seville bookshop, but these had no indications of ground contours. Not all Fiats were equipped with compasses and none had radios. Captain Dequal, the commander of the squadron, himself landed his Fiat fighter on the other side of the Portuguese frontier when trying to reach Cáceres on 31 August, as did at least one other pilot. The Italians had orders to fly west if they got lost, because the Portuguese authorities would return the aircraft and pilots.[31] The Italians complained that the Nationalist command, like the Republicans, used aircraft piecemeal, but so few fighters were actually in service that it is hardly surprising that the beginning of the Spanish Civil War in the air resembled the early days of the First World War rather than what was thought appropriate in the latter 1930s.

Franco's Moors and legionaries were advancing steadily along the valley of the Tagus, called Tajo in Spain. On 23 August they took Navalmoral, and their columns surged 20 miles further west towards their ultimate goal of Madrid. To be nearer the action, Franco moved his headquarters from Seville to Cáceres, to which Bonomi was ordered to transfer all his *Chirris*. The zone was within the reach of enemy aircraft and Franco had more than once been obliged to leave his car when strafed. All possible air support was essential now that the insurgent columns were about to attack Talavera de la Reina, the next large town along the highway to Madrid. Every available *Chirri*, save one which had not been repaired since its heavy landing among the fighting bulls, was flown to Cáceres. On 31 August, three of them met two Dewoitine 372s and a Hawker Fury. Two Italian fighters were shot down. Their pilots bailed out, but while Sergeant

Castellani found his way back to the insurgent lines, Lieutenant Ernesto Monico was captured by Republican troops, who shot him out of hand.

Orders were soon given that enemy pilots captured by the Republican militia should not be harmed but handed over to the authorities for interrogation. Vincenzo Patriarca, the first Italian pilot to be shot down, gave valuable information after his capture on 13 September 1936. His testimony was used by the Spanish government as evidence of Italian pro-Franco intervention.[32] Patriarca was a US citizen who took advantage of an offer for the sons of Italians living abroad to enjoy free flying training in Fascist Italy. More enthusiastic about stunt and acrobatic flying than politically aware, a typical example of the macho and carefree, devil-may-care young flyer of those years, the 18-year-old Patriarca enrolled in the Italian air force to fight in Abyssinia but was promoted, encouraged and well-paid to go to Spain. On 13 September 1936, he was involved in a dogfight and forced to bail out. This was the end of his bravado. Understandably terrified that the Spanish mob would murder him as it had Lieutenant Monico, he was relieved to be sheltered in the US embassy until he was quietly exchanged for a Yugoslav pilot who had been captured flying for the Republic. Eventually, Patriarca returned to Italy but continued to fly.

The Italian *Chirri* fighters were frittered away in individual actions but were replaced by the arrival at Vigo on the night of 26–27 August of the freighter *Aniene* with nine further *Chirris* with their pilots and ground crews, under Lieutenant Dante Olivero, who was killed accidentally before going into action. Though they had sailed through the Strait of Gibraltar, that route was still patrolled by Republican warships, and a Republican warship called on the *Aniene* to stop and identify itself but was seen to turn and sail away rapidly as an Italian cruiser, which probably had been shadowing the freighter, approached. In fact, on 24 August, José Giral, the Spanish Prime Minister, had ordered Republican warships not to enforce the blockade, declared earlier, which Britain in particular had fiercely rejected.[33] The fighters were rapidly loaded on to a train which crawled its way south from Vigo to Seville under an implacable sun.

The S-81 Savoia-Marchetti bombers, concentrated at Tablada air base, from where pleasant escapades could be made to Seville and its exotic delights, were being used for bombing the enemy rear. Among the first of these raids was one on Málaga on 22 August described by the British writers Gerald Brenan and Sir Peter Chalmers-Mitchell.

Gerald Brenan, who would write a number of famous books on Spain, was living in Málaga and recalls in his autobiography:

> One morning at dawn there was an air raid over Málaga. My wife and I stood watching it from our bedroom window when a tremendous explosion from the direction of the port shook the air and a heavy column of smoke poured up into the sky. A bomb had hit the petrol and oil dump that supplied the city [...]

> That evening we listened to the B.B.C. telling us that Málaga had 'probably been completely destroyed'.[34]

Sir Peter Chalmers-Mitchell, who had been Secretary of the Zoological Society of London until retiring to Málaga in 1935, later recalled air raids happening for several days in succession. Probably Sir Peter was writing about the same raid, on 22 August 1936, that was described by Gerald Brenan:

> Then we saw a great cumulus of black smoke rising behind the hill which separated us from central Málaga. We rushed up the hill, and lay panting on the crest. Beyond the harbour entrance a warehouse was blazing, and in front of it and almost touching, rose the volcano [...] our hands over our heads, waiting for the explosion which would shatter central Málaga and even send its crashing reverberations against us. But presently we saw men, like a stream of disturbed ants, at first distractedly running around, and then forming into two thin lines towards the flames, another coming from them, men trying to save the petrol. In actual fact, the heavy oil could not be saved and for several days a column of smoke, turning by night into a pillar of fire, rose perhaps a hundred feet into the air and then, bending in the breeze, swelled into a trail of drifting cloud.[35]

Spain had been neutral in the First World War and had no experience of air raids. Bombing at the beginning of the civil war was very effective and caused severe loss of life until air raid precautions such as blackouts and public shelters were introduced later in the war. Even in Britain, which had been bombed in the First World War and had itself bombed Germany, Prime Minister Stanley Baldwin's often-quoted statement in Parliament in 1932 that 'the bomber will always get through' had contributed to the belief that there was no adequate defence against bombers. The bombings of civilian cities in the early months of the war in Spain were seen in quasi-apocalyptic terms, even though such attacks were usually carried out by one or two aircraft only and with relatively small missiles. Nevertheless, the almost complete absence of air-raid shelters and the lack of advice about how to protect life did lead to significant numbers of deaths and injuries among civilians. In the Málaga bombing on 22 August 1936, a mere three bombers killed fifty-seven people while hospitals were overwhelmed by the injured. In revenge, numerous right-wing prisoners were taken out of prison and executed out of hand.[36]

This bombing and the bombings of concentrations of Republican militia were examples of the tactical use of bombers, not as yet the strategic bombing of cities which the Italian air force would later attempt against Barcelona in 1938 as it put into effect the ideas of Giulio Douhet, the theorist of destruction of the enemy's cities as the ultimate way to achieve triumph in war. Bombing refineries was like destroying railways, roads and bridges, a form of interdicting the enemy's communications and supplies. In time, both the Spanish nationalists and their German allies would develop close air support of ground troops, which required high levels of flying ability and sophisticated means of ground-to-air communication.

On 1 September 1936, Italian aircraft in Franco's service totalled five Savoia-Marchetti S-81s, plus three being repaired. Only one or two *Chirris* were in service. Nine recently arrived fighters were being assembled, four were in Portugal, as yet not returned, and a few others were noted as *fuori uso* or 'unusable' and *non efficienti* or presumably requiring repair. There were eighteen bomber pilots, fourteen fighter pilots, all officers and nine sergeant pilots, presumably of the fighters, plus forty-four other personnel. For the moment, then, Italy was not supporting Franco very enthusiastically.

Bonomi, however, had had enough of his fighters being used in small packets, which was also a constant complaint of the Germans and incidentally also of Andrés García Lacalle, future commander of the Republican fighter arm. After conferring with Franco, who had little idea of the modern use of an air force, Bonomi decided to repair all the fighters available and form them into two squadrons which would fight together. The pilots, provided with appropriate maps, would be flown by the Ju-52 transports along the Seville-Salamanca-Burgos route to familiarize them with the topography of Spain. For this, cooperation with Spanish pilots was necessary, and so on 6 September Captain Joaquín García Morato, an instructor and a future ace, who had flown the German He-51 fighter as well as the Nieuports of the Spanish air force, applied to fly the *Chirri*, in one of which he gave an impressive exhibition of acrobatics over Tablada base. García Morato, of course, had perfect knowledge of the topography of the area over which the Italians were flying.

García Morato's meeting with Colonel Bonomi led to a brief, if amusing, misunderstanding. While the Germans had sent men who had spent the 1920s in Spanish-speaking America training local air forces and had learnt to speak Spanish, the Italians had nonchalantly assumed that Spanish and Italian were more or less the same languages and that there would be no difficulty in mutual understanding. When García Morato saluted Bonomi and formally asked to be permitted to fly the CR-32 *Chirri*, he enquired the name of the squadron leader, to which Bonomi replied 'Dequal'. García Morato understood this as the Spanish '*¿De cuál*? meaning 'Of which one'. 'I mean the leader of my squadron', insisted the Spaniard. 'Dequal', replied the Italian colonel impatiently.[37] It might have been less confusing had Bonomi given the nom-de guerre 'Limonesi' which Captain Dequal had adopted.

García Morato's flight record for 6 and 7 September 1936 refers to fifteen-minute test flights and a combat exercise in the *Chirri*, followed by a training exercise with the squadron.[38] On 8 September 1936, nine *Chirris* were ready on the dry, burnt grass runway of Tablada. The following morning, these nine plus the three available Savoia-Marchetti S-81 bombers took off and flew the one-hour flight to Cáceres, which would be their base for the fierce battles on the Talavera front which it was hoped, would bring the breakthrough to allow Franco's Nationalist insurgents to reach Madrid, thus ending the war.

In the Balearics

In the meantime, however, Italian interests had brought their aircraft to the Balearic island of Majorca. At first, the garrisons of Majorca and Minorca, the latter with its valuable naval base of Mahón, had risen against the Republic, but the non-commissioned officers at Mahón had counter-rebelled, and Minorca would remain in Republican hands until the end of the war. On Majorca there were a few aircraft, but none was in a condition to fly. Thus, the few minor attacks on the island by Republican Savoia-62 seaplanes from Minorca and a Dornier Wal from Barcelona, while having no significant effect on the insurgents, terrified the population. Apprehensive of an attack from the sea, the insurgents decided to try to organize their own means of defence. On 2 August, two representatives were taken to Italy on a German ship. The Italian authorities recognized that not only was it easy to supply war material to Majorca but also that the island, with its harbour and seaplane base, would be an excellent base for naval and air action on the Spanish mainland.

However, there would be a cost. A large proportion of this was contributed by the Majorca-born Spanish multimillionaire Juan March through the Kleinwort-Benson Bank in London.

In March 1936, acting for March, Kleinwort's had made half a million pounds available to the conspirators, a sum which was increased at the outbreak of the war to £800,000 and in December 1936 to £942,000.[39] March also paid for the Savoia-Marchetti S-81s and for the charter of the Dragon Rapide which had flown to the Canaries from England to carry Franco swiftly to Morocco on 18–19 July 1936.[40] In July 1936, March deposited 178 tons of gold bullion in the Banco di Italia.[41] About a ton of gold was collected on the island, in a system of 'suggesting' sums that the wealthy could contribute, while the Bank of Spain's local branch was more or less obliged to hand over sterling and French francs. Italian seaplanes would fly to Majorca on 17 August and be followed by a ship with six fighters and three antiaircraft batteries.

In the meantime, by 9 August a chaotically organized expedition from Barcelona and Valencia had recovered the smaller islands of Formentera and Ibiza for the Republic, and on 13 August four large transports escorted by ships of the Republican navy appeared off the coast of Majorca. This mostly militia force of a mixture of between six and seven thousand men did not disembark until 16 August. Catastrophically for the Spanish Republic, the indecision, general incompetence and rivalry between the commanders led to stalemate. Only a small part of Majorca was occupied and there was no obvious plan of action.

Three Italian Savoia S-55X seaplanes, of the type made famous by a transatlantic flight led by the future Marshal Balbo in 1933, reached Puerto Pollensa in the north of the island on 19 August, refuelled, took off again, attacking and dispersing the Republican warships as well as bombing concentrations of militia. When, however, one Italian seaplane was later destroyed by a bomb launched by aircraft from Barcelona, the other two seaplanes returned to their base. After further appeals transmitted through Italian warships at anchor off Palma, capital of the island, Mussolini sent a murderous and theatrical Fascist, Arconovaldi Bonaccorsi, to ginger up the resistance of the insurgents. Arriving on 26 August, and adopting the name of 'Count Rossi', he strode around in high black boots and a large white cross round his neck, his black uniform hung about with pistols, grenades and daggers. At last, on 27 August, a ship arrived with three *Chirris* and three Macchi-41 single-seat biplane seaplane fighter bombers, under the command of Major Gallo, a dozen anti-aircraft guns, ammunition and bombs. When three S-81 bombers arrived, the presence of several modern Italian aircraft persuaded the Republican command to begin to get ready to evacuate the landing force.

Bonaccorsi and his small staff immediately set about creating a modern air force base, with hangars, roads, suitable fuel and ammunition stores, accommodation and telephone lines. Observers were posted at fourteen lookout posts. On Sunday, 30 August, the middle class of Palma, greatly relieved at the removal of the threat of the Barcelona militia, heard Mass in the cathedral. Present were Italian pilots in Legion uniform, who marched out at the end of the service between lines of applauding people, as a glittering pair of *Chirris* performed acrobatics overhead.[42] Over the next days the S-81 bombers bombed concentrations of militia as well as the Republic transports and warships at will, untouched by the anti-aircraft fire from the vessels which left finally on 3–4 September. As the militia were re-embarked, they abandoned unburied corpses, weapons, vehicles and war material of all sorts. There would be no mercy for prisoners. While the Italian Major Gallo was outraged at the mass killings of captured militia, on 6 September, Bonaccorsi, swaggering on his white horse, led a victory parade through the streets of Palma at the head of his murderous gang of 'Dragoons of Death'.

The Republican government had made its decision. Desperately gathering all its resources to defend Madrid, it would abandon Majorca, and the smaller islands of Formentera, Ibiza and Cabrera, soon reoccupied by the insurgents, leaving only Minorca in Republican hands. This was a strategic error as great as allowing Franco's forces to cross the Strait of Gibraltar, because Majorca would within a few months become a major surface and submarine base, from which for the rest of the war, aircraft would harass traffic making for the Republican ports of Barcelona, Tarragona, Valencia, Alicante and others along the Mediterranean coast.[43]

On 3 September, as the militias were abandoning Majorca, the Moors and legionaries of Franco's expeditionary force took Talavera de la Reina, 76 miles south-west of Madrid. On 4 September, a new Republican government was formed, led by the socialist Francisco Largo Caballero, who also held the portfolio for War but established a new ministry – of the navy and air – under the likewise socialist Indalecio Prieto. Every aircraft that was fit to fly was put into the air to try to reoccupy Talavera. The Bréguet XIXs and Nieuport 51s, however, were no match in most circumstances for the *Chirris*, whose squadron was now known by the nickname *La Cucaracha* or the 'beetle', as in the well-known Mexican song. On 11 September *La Cucaracha*, led by García Morato and Dequal, downed nine of its opponents. By mid-September, fewer and fewer Republican aircraft could be put up, even though accidents and losses had reduced the number of CR-32s to about ten.[44] But most of the French machines which had been supplied to the Republic had been lost, and the latter had no more sources of aircraft, unlike the insurgents, who could count on reinforcements from Germany and Italy. Only three or four fighters were flying for the Republic by 23 September. In contrast, more aircraft had arrived from Italy. These were IMAM Romeo 37 bis, highly manoeuvrable, two-seater reconnaissance biplanes, with a maximum speed of 205 mph, with radio and an aerial camera, armed with twelve small bombs on racks under the wings and three 7.7 machine guns. On 29 September, the freighter *Aniene* brought ten of these to Vigo and a further squadron of twelve *Chirris*. Taken by train to Tablada, they were rapidly assembled into two small squadrons, one Spanish-manned. Known in Spain as *Los Romeos*, they went into action over Madrid on 23 October, destroying two observation balloons floating over the extensive park to the west of the capital called the Casa del Campo. A further ten *Romeos* were unloaded at Cádiz, three of which were damaged in test flights. By 1 October 1936, forty-seven Italian aircraft were flying in Spain, mostly with Italian professional military pilots. There were twenty-four Fiat CR-32 *Chirris*, fifteen *Romeos* and the original eight S-81 bombers, one of which had blown up accidentally at Tablada. Four *Chirris* had been lost in combat and five in accidents.

Between November 1936 and January 1937, an additional sixty Fiat CR-32s arrived in Spain (thus making the total so far supplied 120). Among the fighters supplied were twenty examples of the CR.32 bis *quadriarmi* (four-gun). These machines were heavily armed with two 12.7mm guns fitted above the forward fuselage and synchronized to fire through the propeller arc, as well as a pair of 7.7mm guns housed within the lower wings. Apart from the ground crews, thirty-three fighter pilots arrived under Lieutenant Colonel Alberto Canaveri (alias

'Franco Signorelli'), who was to take over the fighter wing of the Italian air force in Spain.

By the end of 1936, the CR-32 *Chirris*, particularly when piloted by Spanish aces such as Angel Salas and Joaquín García Morato, had shown themselves capable of tackling all the aircraft that the USSR had sent starting in October 1936.

Reflecting the escalation in the conflict and probably because there was no longer any secret about Italian intervention in Spain, the *Aviación del Tercio* was dissolved on 28 December 1936 and replaced by the *Aviazione Legionaria*. The latter combined the majority of the aircraft that had been sent to Spain from Italy together with all Italian airmen participating in the mission, as well as some Spanish air crew. An autonomous entity, the *Aviazione Legionaria delle Baleari*, on Majorca, was equipped with bombers intended for anti-maritime or coastal operations, while, during December, three more CR-32s joined the *Squadriglia Mussolini* at Palma.

In early December, Italy also organized the *Corpo Truppe Volontarie* (Volunteer Troop Corps) made up of infantry divisions, arms and vehicles, and it also mobilized the merchant fleet and provided escorts from the Italian navy.[45]

Chapter 4
France sends aircraft: Léon Blum's dilemma

Having lost control of the situation and faced by irresistible demands for arms to be distributed to allow the people at large to resist the military coup, the Prime Minister of the Spanish Republic, Santiago Casares Quiroga, resigned. On the night of 18–19 July 1936, the centrist Republican Union Party politician Diego Martínez Barrio formed a government and appealed to the insurgent General Mola in Pamplona to call off the military uprising. His appeal was in vain. The Minister for the Navy, José Giral, then took over the premiership and, late on Sunday, 19 July, sent a telegram to Léon Blum, the French Prime Minister, who read it when he came into his office the next morning.

> Have been surprised by dangerous military coup, Request you come to agreement with us immediately regarding supply of arms, aircraft. Fraternally Yours, Giral.[1]

Blum had been in power only since 5 June of that year. He was now faced with what was to become a deep crisis and cause of profound schism in French opinion. The victory of the Popular Front reflected the desire to protect the French Republic against the assaults of the extreme right-wing movements which had nearly overthrown it in riots on 6 February 1934. Not surprisingly, Blum's reaction to the appeal of the Spanish Prime Minister of a Popular Front government threatened by a right-wing uprising was to send armaments immediately. The next day, Tuesday, 21 July, two Spanish air force officers, Majors Juan Aboal and Ismael Warleta, flew to Paris with a detailed list of requirements. These were for eight 75mm field guns, fifty machine guns and twelve million cartridges, one thousand rifles, one million rifle cartridges, twenty thousand bombs and twenty Potez 540 bombers with crews.[2]

The Potez 540 was a twin-engine, single-wing bomber and reconnaissance aircraft with a crew of five, in service with the French air force. It carried one ton of bombs and was well-equipped with machine guns, radio and oxygen. However, only four machines were available to be sent to Spain. Furthermore, the French Foreign Minister, Yvon Delbos, was anxious that they should not be ferried to Spain by French pilots for fear of international repercussions. On Wednesday, 22 July, Delbos and Blum went to London for a previously arranged conference on Anglo-French policy. Although Spain was not on the agenda, Anthony Eden, the British Foreign Secretary, and probably also Sir Samuel Hoare, the First Lord of the Admiralty, gave strong hints to Blum that sending arms to the Spanish Republic was not approved of in London. In the meantime, the Spanish ambassador in Paris, Juan de Cárdenas, sympathetic to the insurgents and about to resign, warned British diplomats of the French intention to arm what he saw as dangerous revolutionaries in Spain. While Pierre Cot, the French Minister of Aviation and highly sympathetic to the Spanish Republic, wanted to dispatch aircraft to Spain there and then, the French Foreign Ministry, apprehensive of German, Italian and above all British hostility to such a move, tried to forestall him. When Blum and Delbos returned to Paris on the evening of Friday, 24 July, at about the same time that Franco's emissaries were landing in Berlin, they discovered that the right-wing French press had been informed of Blum's intention to respond positively to Giral's appeal by the military attaché of the Spanish embassy in Paris, Colonel Barroso, who would later become Army Minister in Franco's government. Moreover, the right-wing newspapers were attacking Blum violently. The French premier learned to his dismay that not only were the presidents of the Republic and of both chambers of the French parliament uneasy, but not even all the members of his own government were in agreement with his intention to arm the Spanish Republic.

The Spanish socialist and ex-Minister of Justice, Professor Fernando de los Ríos, who was by chance in Geneva, was asked by the new Spanish Prime Minister, Giral, to go hotfoot to Paris and serve as temporary Spanish ambassador after Cárdenas's resignation. On 24 July, de los Ríos presented the Spanish official requests for aircraft, but he had as yet no diplomatic status, which gave the French Foreign Ministry, always referred to at the time as the *Quai d'Orsay*, as well as the unwilling Potez management, opportunities to create delays in supplying the aircraft. The *Quai d'Orsay* also refused to permit French pilots to ferry planes to Spain, despite the lack of Spanish pilots with the necessary experience and ability.

The French cabinet met the following Saturday afternoon, 25 July, at 4 pm. In their minds was the fear that, by intervening in the Spanish conflict, they risked losing the support of Britain in the possible major European war which was the nightmare of French politicians and which might be ignited by the competing interests of the great powers in Spain. Blum himself wanted to resign but was persuaded by the new Spanish ambassador to stay. As the French ministers argued, a Spanish DC-2 airliner was landing at Le Bourget airport with eighteen cases containing a down payment of £144,000 in British sovereigns (most of the Spanish gold reserve was held in coins). The first three transfers of gold from the Bank of Spain to Paris totalled £432,000 which was exchanged for 53 million paper francs.[3]

In the end, the French government decided not to send aid officially to Spain but to tolerate 'private' arrangements and to allow unarmed aircraft to fly to Spain. This was in marked contrast to what was happening, that Saturday night, 25 July, in Germany, unknown of course in Paris, when Hitler decided to send Junkers transports and Heinkel fighters to the military insurgents. Thus the Spanish Republic was already being abandoned by the Democracies while Hitler was being extremely helpful to the insurgents and particularly to Franco. When news broke on 30 July that Italy also had sent aircraft to the insurgents in Spanish Morocco, the French opposition blamed Blum and Cot for 'provoking' Mussolini by agreeing to send arms, though in fact no French aircraft had left for Spain as yet.

Blum's government then proposed that all European countries should be invited to agree to refrain from sending armaments to either side in Spain, but while waiting for agreement, Blum insisted that France would do what it thought fit. Military aircraft would be flown to the Spanish Republic, but the sale would be negotiated through commercial agencies. In any case, the aircraft must go unarmed.

The matter was complicated, because some of the aircraft first offered to Spain were not fit to be used in action. In the end, six Potez 540 bombers were made available. They were flown unarmed from Toulouse to Barcelona on 8 August 1936. The Potez was supposed to have five rapid-fire Darne machine guns, but in action in Spain it was defended by a couple of older Lewis guns only, which made it an easy target for fighters and thus led to its condemnation as a 'flying coffin' or, as it was known in Spain, *ataúd volante*. That was not all. The Potez bombers went to Spain without bomb racks or bomb sights, all of which had to be improvised by Spanish mechanics. Further, Potez aircraft were sent in 1936 and some in 1937,[4] all as commercial transactions. The supply of German aircraft to Franco was also masked as a commercial arrangement, but it was done by the personal order of Hitler, in contrast to the French supply of aircraft to the Spanish Republic, which was the result of an uneasy compromise which reflected sharp differences of opinion within a democracy.

Attempts to find French aircraft to be sent to Spain were blocked, in some cases by the unwillingness of the manufacturers to divert orders, in others by leaks to the right-wing press. In the end, all that became immediately available were a number of Dewoitine 372 fighters, from a cancelled order originally intended for Lithuania. On 4 August 1936 reserve sergeant pilots and other volunteers flew six Dewoitines from the manufacturers' airfield to Toulouse, from where the flight to Barcelona was within the fighters' range. Other Dewoitines were flown to Toulouse over the next two days. One pilot lost his bearings, landed in a field, breaking the undercarriage and a wing.

As the Dewoitine fighters and Potez bombers waited at Toulouse, Jean Moulin, *Chef de Cabinet* in the French Air Ministry, overruled French customs and police officials who wanted to prohibit the planes flying to Barcelona. Sir George Clerk, the British ambassador, did his best to persuade Yvon Delbos, the Foreign Minister, to prevent the planes leaving. On 7 August, one pilot, retired Sergeant Blois, flew to the Catalan capital where at 7.20 pm he landed the first French aircraft to be sent to the Spanish Republic.

During the lengthy and stormy French cabinet meeting on Saturday, 8 August, as it became evident that Blum saw no alternative to the pan-European agreement not to supply arms to Spain, to be known as *Non-Intervention*, if he was to maintain the support of Britain, the planes took off for the hour's flight to Barcelona.[5] Three of the Dewoitines were damaged on landing, which shows how difficult they were to fly for pilots who had not undergone a period of instruction. Altogether, fourteen Dewoitines went to Spain in 1936 and possibly a total of twenty-six throughout the war.[6]

Republican Spain thus had to buy aircraft where it could and at very high prices.[7] There was, in any case, no reasonable comparison between the French machines and those sent by Germany and Italy. Though the Dewoitines were of high fighting quality and enjoyed an unmatched speed of climb, they arrived without much of the equipment needed for them to be offensive weapons and were tricky to land for untrained pilots. This was well-known, for the US cruiser *Quincy*, in Barcelona harbour to repatriate US citizens, reported, 'Planes believed to require expert handling and Spaniards not believed capable.'[8]

Spaniards, however, were not flying them for the moment, but the largely reserve French pilots, who had not been trained to do so either, damaged some of them on landing. When the squadron was moved later to Cuatro Vientos, near Madrid, two more Dewoitines crashed on landing, leaving eleven.

Absence of combat equipment and the inexperience of crews were precisely the differences between the French and the German and Italian aircraft, which came with all their equipment and were manned by Luftwaffe and Italian air force military pilots, gunners and bomb aimers who knew their planes.

As the Spanish pilots and mechanics at El Prat, Barcelona's civil airport today, examined the French aircraft, they were taken aback to see that not only were the aircraft not armed but neither were there any mounts or any way of installing guns. The electrical firing mechanisms, the synchronization mechanisms for the guns which were intended to fire between the rotating propeller blades, the ammunition boxes, gun mounts and sights had all been removed, if in fact they had ever been present, given that the planes were not taken from squadrons but had come from the factory. The Dewoitines had been intended for Lithuania, so the various instructions were in Lithuanian. A day or two later, a crew of Dewoitine mechanics was flown out to repair the damaged machines, but the essential equipment never arrived, and Spanish mechanics had to rig up unreliable contraptions in order to install machine guns. The Dewoitine was superior to the Heinkel-51 and probably to the Fiat CR-32 fighters which the insurgents were receiving but only with properly trained pilots and appropriate armament. Nevertheless, the first two Dewoitine 372s to be armed in an improvised way, accompanied by a Hawker Fury, managed to down two Italian fighters on 31 August. Unfortunately, the success was not repeated. Four Dewoitines were soon shot down, but the others were destroyed by bombing and in accidents. A few were repaired and survived until the end of the war.

The Potez 540 bombers, of which fourteen were flown to Spain in 1936 and four in October 1937,[9] provoked contradictory comments. Their official cruising speed was 150 mph, but this was at a height of fifteen thousand feet. Since the primitive bombsights which had to be improvised for them in Spain, together with the lack of suitable maps, required the Potez to fly at three thousand feet or lower if the bomb aimers were to locate their targets and have some chance of hitting them, the maximum speed at which they could fly was about 100 mph.[10] The scarcity of fighters to escort the Potez added to its reputation as a 'flying coffin'. Again, the Potez could not defend itself even if it were armed with all the machine guns it ought to have, against an experienced fighter pilot in a Nieuport. Nevertheless, according to the Soviet pilot Lieutenant Sharov, who flew the Potez and then the fast Russian Tupolev SB2 bomber, the Potez had a better field of fire than the SB-2 and was easier to control, but its major disadvantage was its low speed. Furthermore, the crews were multilingual and had not flown together before, so they cannot be properly compared with the Germans and Italians who had been trained to fly the aircraft they manned and had often been in the same squadrons. Similarly, the four Marcel Bloch 210 bombers which were sent from France to Spain at the end of October 1936 were immobilized for a long time with engine faults. Their armament, as with all the French aircraft, was improvised and limited to a Lewis gun.[11]

The Malraux squadron

The end of the First World War, together with the uncertain economic conditions of the interwar years, released eager and unsettled young men on to the flying market. Some of these, from the United States, from Britain and from France, among other countries, found their way to Spain. Because it was Spain's neighbour and given the willingness of at least part of the French government to supply aircraft to the Spanish Republic, French reserve pilots ferried planes to Barcelona and others flew them in battle. One unit, known as the *Escadrille Espagne* or 'Spanish Squadron' and later as the 'Malraux Squadron', took its name from its founder and organizer, the French writer André Malraux.

Soon after Giral's appeal to Blum, Malraux appeared on the scene. In his mid-thirties, he was the epitome of the politically committed, anti-fascist intellectual and a sympathizer with communism. He had led protest movements for the release of Georgi Dimitrov and Ernst Thaelmann, prisoners of the Nazis, as well as for the writer Ludwig Renn (pseudonym of Arnold Vieth von Golssenau) who commanded the German-speaking Thaelmann battalion of the XII International Brigade in the Spanish Civil War. Famous for his prize-winning novel *La Condition Humaine*, thirty years later Malraux would be Minister for Cultural Affairs in the de Gaulle government of 1958–68.[12]

Malraux had been in Spain in May 1936, visiting the leaders of the Popular Front and being welcomed by progressive intellectuals, including the poet Federico García Lorca. On his return, he had lectured on the Spanish situation.[13] On 24 July 1936, at the request of the Air Minister, Pierre Cot, Malraux went to Spain to report on the air war. He and his wife, Clara, flying in a private Lockheed Orion airliner owned by the French Air Ministry, touched down at Barajas airport near Madrid early the next morning.[14] Malraux spoke to senior officers in Madrid and flew later to Barcelona. He must have been told that there was a shortage of pilots and that more modern aircraft than those in service in Spain were needed, for, convinced that the best thing to do was to raise a squadron of aircraft and recruit volunteers, he returned to France on Monday, 27 July. On 30 July, Malraux spoke at a great protest meeting about Spain held at the Salle Wagram in Paris. The general understanding in left-wing circles in France was that aircraft were the most important weapon that the Spanish Republic needed, and consequentially the crowd chanted rythmically *Des avions pour l'Espagne*! together with the First World War slogan *Ils ne passeront pas*!, first heard in the battle of Verdun in 1916 and which would be copied when Madrid was under siege as *No pasarán*! Fernando de los Ríos and Alvaro de Albornoz, who succeeded him as Spanish ambassador in Paris, asked Malraux to be nominal commander of the squadron of volunteers that he, together with Pierre Cot and his staff, was to recruit. Malraux worked with a number of sympathizers including many great pilots of the 1920s, led by Jean Moulin, in charge of civil aviation in the Air Ministry and a later coordinator of the anti-German resistance, who would be murdered by the Nazis in 1943. Their aim was to outwit the obstruction of officials who wanted to prevent any intervention by French citizens in the Spanish war. Pilots were engaged by the Spanish embassy in Paris. Some were older aviators from the Great War;[15] others had fought against the Japanese in China, while others were refugees from Nazism and Fascism. Some had flown more recently in the 1932–5 Chaco war between Bolivia and Paraguay, and others had flown for the Emperor Haile Selassie in defence of Abyssinia against the Italian attack of

1935–6. Many had to be rejected because of their age or for medical reasons. The pay offered was generous. It would be, for 1 month, to be renewed if both parties agreed, 50,000 francs, with a life insurance policy of 200,000 francs. The contracts bore the signature of Colonel Camacho, the Undersecretary for Air in the Spanish government.[16]

In order to give the operation an official character, the Spanish Republican War Ministry authorities gave André Malraux the rank of lieutenant colonel. This title gave him authority as leader of what became known as the *Escadrille Espagne*, which flew the Dewoitine fighters and the Potez bombers and sundry other machines which had been bought from the French manufacturers. Malraux helped to hire crews for the planes, mainly volunteers and professional civil pilots. Some were idealists; others went for the huge sums of money that could be accumulated in a few months, assuming the pilot survived. Still others were adventurers in love with flying.

The volunteers went to the cinema in Toulouse the night before they flew the aircraft to Barcelona. They watched a film about the war in Spain in which Spanish Bréguet XIXs were shown being attacked by German He-51 fighters.[17]

Those who ferried the French aircraft from Toulouse to Barcelona the next day were listed by the airport inspector at Toulouse.[18] Two or three were test pilots but most of them were reservists who probably had little or no experience of flying the up-to-date Dewoitine. Four of the pilots who flew planes to Toulouse also flew them on to Barcelona. Two of these were well-known and flew missions in Spain. One was Victor Veniel, a 32-year-old reserve French air force captain, who used the nom-de-guerre of Valbert, though he was known to the squadron as 'Capitaine Vic'.[19] After service in the First World War as an artillery officer, he became so successful as salesman for the Potez aircraft company that he earned a small plane as commission. He learned to fly it and then joined the French air force reserve. Having provided some very efficient services for Pierre Cot, the Minister for the Air, when the latter resolved to send aircraft to Spain, his top civil servant, Jean Moulin, immediately summoned Veniel and charged him to report on the nature of air operations in the Spanish war. He would later fly in the Second World War, be decorated and promoted to colonel. The other was Jean Dary[20], a 41-year-old First World War ace who shot down at least four insurgent machines and was invited later by the Russians to try out one of their state-of the-art I-15 fighters, though they may have suspected him of being what he was, an agent instructed to send reports back to Pierre Cot.[21] Dary, in the view of the Spanish fighter pilot Andrés García Lacalle, who knew many of these men well, was the most responsible and courageous of all the Malraux squadron. He also had a criminal background, having served a gaol sentence for disguising stolen cars for resale. Another, Abel Guidez, a 28-year-old civil pilot, found his métier in Spain as a fighter pilot, scoring ten victories and being selected to command the squadron after Malraux was relieved at the end of 1936.[22] It was Guidez who flew the Soviet journalist Mikhail Koltsov to Barcelona on 8 August, probably in one of the Potez bombers since Koltsov was accompanied by two other passengers. The Russian journalist gives a vivid picture of Malraux at Toulouse as the planes flew off, 'standing on the runway, his legs apart, his hands in his pockets and a cigarette in his mouth, for all the world like a music-hall manager at a general rehearsal'.[23]

Other volunteers included Adrien Metheron; Hantz, an Alsatian who had flown in the German air force in the First World War; Raymond Maréchal, who would be Malraux's main ally in the French resistance in 1944; and the Belgian Heilman, who at the age of sixteen had seen the German invasion of his country and had escaped in 1917 to join the French army where he had become a pilot. Later, he had fought in Poland and in China. With them were the Argentinian Jean Bélaïdi; the Parisian primary school teacher Labitte; the Italian Giordano Vizzoli; the Belgian Paul Nothomb; the Palestinian Jew Yehezkel Pikar; Henri Gensous; 23-year-old Michel Bernay, the youngest of the squadron, who had only recently left the French air force; and the oldest, 45-year-old François Bourgeois, a First World War ace who had smuggled alcohol during the US Prohibition years and taken part in revolutions in the Argentine. The other pilots called him *Dieu le Père* or 'God the Father' because of the halo-like circle of white hair which surrounded his mostly bald head. The squadron began with seventeen pilots altogether, accompanied by a few mechanics.[24] Thus, not all the crew of a Potez would be able to speak and understand orders given in French over the noise of the engines, even with earphones, and even fewer could handle Spanish. None would have had experience of combat unless they had fought in the First World War. It is true that the Germans who flew for the insurgents did not have war experience either and the Italian experience was limited to the practically defenceless Abyssinians, but the Germans and Italians were highly trained and disciplined military pilots. Furthermore, without the machine guns with which it ought to have been armed, the Potez was very vulnerable to fighters. In a later article in a professional journal, Dary suggested that bombers could be defended only if they flew in very tight formations ('le vol de groupe très serré').[25]

For that first week in chaotic Barcelona, a city in the hands of its working class, as George Orwell vividly describes it in his *Homage to Catalonia*, the pilots were applauded wherever they went in the confiscated Rolls-Royce limousines which they were allotted.[26] To avoid word getting out that the Spanish Republic had contracted foreign pilots, the *Escuadrilla España*, as it was known in Spain, was limited to bombing and reconnaissance missions and ordered to avoid combat lest the pilots be shot down and undergo interrogation. However, it was not long before the squadron was engaged in combat.

After the pilots and the planes arrived at Madrid in August 1936, and were given rooms in two of the best hotels, the Florida and the Gran Vía, Malraux himself took charge of the organization of the squadron. He was not a pilot, but he did fly on some of the twenty-three missions that the squadron carried out. While waiting for the Dewoitines to be armed, Valbert and the others flew missions in whatever aircraft came to hand, including an aged Latécoère from the mail plane service of the 1920s.[27] The squadron flew its first sorties in August 1936, scoring its first victory when Jean Dary and another French pilot shot down two Italian aircraft.[28] In the intense summer heat, the insurgent forces of Franco's expeditionary force were rapidly advancing through Extremadura on their way north to join the forces of General Mola advancing southward from their bases in northern and north-western Spain. On 15 August 1936, three bombers from Malraux's *Escuadrilla España* joined three other Republican bombers in an attack on an insurgent column. Now, for the first time in the Spanish war, pilots from other countries saw each other. The French airmen spotted six German Ju-52s but avoided combat.[29] On 16 August 1936, a Fiat CR-32 fighter shot down one of the fighters in Malraux's squadron. The pilot, surnamed Thomas, parachuted into the Tagus River. He knew that foreigners were always suspect, for he had often been held up in the streets of Madrid and questioned aggressively by self-appointed Republican security patrols who were always on the lookout for insurgent agents. He swam to the river bank but then was perplexed. Were the bellicose Spaniards he saw Republicans or Francoists? Which salute should he give, the clenched fist or the Roman outstretched arm? He compromised with a timid hands up gesture as the river water dripped off him. When he realized he was in the Republican zone, he clenched his right fist. Once he had convinced them that he was a French volunteer for the Republic, they carried him in triumph to the nearest village. He was entertained in successive villages over the next two days before he found his way back to Madrid. On the same day, several bombers from the *Escuadrilla España* bombed the recently conquered city of Badajoz. In other action on 15 August, two fighters escorted two Potez 540 bombers in a raid over Teruel, on the Aragon front. When three Italian Fiats attacked the formation, the two fighters of Malraux's squadron engaged them in a desperate air battle. The youngest pilot in the *Escuadrilla España*, Michel Bernay, shot down one of the Italians, while Victor Valbert pursued one of the others.[30]

On 16 August 1936,[31] the squadron received a message that an insurgent motorized column under Major Carlos Asensio Cabanillas, following the Guadiana river eastward from Mérida, was massed in and around Medellín. As the Potez planes flew over the town, they saw that the plaza was jammed with trucks and other vehicles, completely exposed to attack. The pilots of two Potez bombers and a DC-2, chancing that the opposing forces did not have anti-aircraft weapons, descended to the dangerously low altitude of one thousand feet before releasing their bombs. The attack virtually wiped out part of the column.[32] Three days later, however, the insurgents bombed the important base of Cuatro Vientos, to the south of Madrid, damaging several of the Dewoitine fighters of the *Escuadrilla España*. Few were left in operative condition.

Two of the Potez bombers, with the humorous names of *¡Aquí te espero!* ('I'm waiting for you here') and *¡Voy corriendo!* ('I'm hurrying!') carried out some long-distance raids, with doubtful success, against the insurgent base at El Ferrol on Spain's north coast and along today's tourist Costa el Sol in the south.[33] On 25 August, however, *¡Aquí te espero!* was shot down by the Spanish insurgent Captain Angel Salas Larrazábal, who had finished his pilot's training in 1929 in top place and had completed 1,625 flying hours before the civil war in which he would become one of the insurgent aces. In July 1936, he had taken an aircraft and flown from the loyal base at Getafe to Pamplona to join the military uprising. He would end his career forty years later with the distinction of Captain General and be appointed a member of the Regency Council. The Potez *¡Aquí te espero!* which Salas downed was flown by Captain Joaquín Mellado, who was a director of the Spanish civil airline *Líneas Aéreas Postales Españolas*. It was Mellado who, on 18 July 1936, had flown one of the airlines, Douglas DC-2s, to bomb the aerodrome at Larache in Spanish Morocco. *¡Aquí te espero¡* was not escorted by fighters and was forced down in Republican territory but Mellado and his co-pilot mistook the civilians who came towards them for Francoist militia and shot themselves rather than risk capture.

On 1 September, the Malraux squadron bombed an insurgent airstrip near Valladolid. During the rest of the month, the squadron concentrated on unsuccessfully bombing the thick walls of the fortress or Alcázar of Toledo, where about 1,200 insurgent *Guardias Civiles*, with women and children and also hostages, had been holding out against besieging Republican militia and miners from the Asturias coalfields who had been brought south to use their dynamiting skills against the immensely thick walls of the Alcázar.

The Russian journalist Mikhail Koltsov describes how the French pilot Abel Guidez had to fly bombers and fighters almost without rest. One day Guidez was flying one of three fighters escorting a bomber. At Barajas, they were waiting anxiously for the aircraft to return. Guidez was the first to land. He reported that the expedition had been attacked by nine Fiat CR-32 fighters. Two Dewoitines had escaped but the Potez, flown by a Spanish pilot named Gustavo, was limping back. At last, the Potez landed. Gustavo was wounded in several places, as were the bomb aimer and the machine gunner. The aircraft needed extensive repair. Gustavo was taken to the military hospital and later died.[34]

By early October, however, only six aircraft and fifteen pilots were left in Malraux's *Escuadrilla España*.[35] Despite Dary's insistence that all six should be used together, the Spanish command sent them off in groups of three to protect its bombers. In the second fortnight of October, Getafe was occupied by the insurgents, and one Dewoitine, one Fury and one other plane were moved to the Republican airfield at Alcalá de Henares, east of Madrid. The fighters were lost one by one so that in the end only the single Hawker Fury remained. Meanwhile, several French pilots had been transferred to Barcelona, where they were able to fly two (of four which were sent but crashed in the hands of inexperienced Spanish pilots) Gourdou-Lesseure single-seat fighters, of less value than the Dewoitines, some of which had at last been armed, though with Vickers machine guns which frequently jammed.[36]

In all, twenty-six modern French military aircraft were flown to Republican Spain without armament or the means to install it. In addition, some individual older military aircraft were delivered, again unarmed, as well as some small civilian planes. The most recent and trustworthy evidence suggests that a maximum of sixty-two aircraft of all types went from France to Spain, of which twenty-five played some part in the fighting after long delays to fit them with adequate armament. Missing spare parts, from magnetos to undercarriage wheels, kept planes on the ground for weeks.[37]

The senior officers of the Republican air force developed an aversion towards the *Escuadrilla España*. In particular, Colonel Camacho made a lengthy complaint to the air force commander, Ignacio Hidalgo de Cisneros. After citing several instances of indiscipline, Camacho wrote:

> All the above shows that the French personnel behave not merely with autonomy but with absolute personal independence close to anarchy, which means that not only are they not useful but they are indeed disruptive. They bring disorder wherever they go and they cannot be trusted to carry out the missions assigned to them.
>
> Thus the undersigned is of the view that these men should be kept together on one airfield and under the orders of a person who has authority over them and takes responsibility.[38]

Hidalgo de Cisneros wrote later that Malraux had no experience of flying and could not function efficiently in command of a squadron. As for the pilots, apart from a few genuine antifascists who came to Spain to fight for their ideals, the others were mere adventurers who came for the money they were paid. They did nothing useful and caused many problems.[39]

For their part, the pilots of the squadron felt often that they were not properly employed by the Spanish air command and that the latter to some extent lacked the discipline and order that the situation demanded. Given that the older French pilots had flown in the First World War and that the others had been in the French air force, the tension may have reflected different national cultural behaviour. Malraux and the pilots of the *Escuadrilla España* did indeed live independently and luxuriously at the Hotel Florida in Madrid, along with foreign journalists, guests of the government and adventurers from various countries.[40] Colonel Hidalgo de Cisneros tried to have the Malraux squadron expelled, but the government insisted that expulsion would have a bad effect on French opinion.[41] However, the mercenaries found, at the end of November 1936, that their contracts were not to be renewed. After that, the squadron was composed of true volunteers.[42]Many of these came from the International Brigades' base at Albacete, where Malraux recruited men who claimed to have served in an air force.[43] It was at this point that, to mark a new beginning, Paul Nothomb, who later used the name Julien Segnaire (echoing the airfield of La Señera where the squadron was based), suggested to the Spanish Air Ministry that the squadron should be renamed *Escuadrilla Malraux*. Nothomb became the political commissar, a post which existed in all units of the Republican forces and was intended to ease relations between the men and the officers, particularly those who were professionals and were often suspected of disloyalty.[44]

Colonel Camacho's hostile report may explain why the *Escuadrilla España* was sent away from Madrid to La Señera near Chiva in Valencia. Here, for three months over the winter of 1936–7 they bombed the insurgent salient at Teruel in Aragon. Potez aircraft were steadily lost, without replacement. The Malraux squadron suffered a chronic shortage of spare parts and supplies which reduced the number of planes available for missions.

In one of the Malraux squadron's last operations, on 11 February 1937, at 10 am, a Potez bombed an enemy column which was advancing behind the mass of refugees fleeing along the coast from Málaga to Almería. Attacked by Fiat CR-32 Italian fighters, the Potez, with its crew wounded and its engines destroyed, ditched off Motril in shallow water. The unwounded members of the crew carried the wounded men to the beach and the road, among hundreds of ragged and footsore refugees fleeing the massacres taking place in Málaga. Finally, they found the Canadian doctor Norman Bethune, who had brought a blood transfusion service to Republican Spain. His ambulance took the wounded airmen to Almería.[45]

One of the Potez planes was shot down on 27 December 1936 over a mountain range near Teruel. The episode inspired André Malraux to write his 1937 novel *L'Espoir*, translated into English as *Man's Hope*. It was made into a film called *L'Espoir, Sierra de Teruel*, in 1938–9, but not released until 1945.

So many men had been killed or seriously wounded, and so many aircraft lost, that there was no point in continuing. Finally, the Malraux squadron was dissolved in February 1937 and its remaining aircraft incorporated into the Republican Air Force.

Paul Nothomb, alias Julien Ségnaire, the commissar of the Malraux squadron, wrote a novel which, though by its very nature a fiction, includes a passage which brings a bombing raid vividly to life. One of the DC-2s of the Spanish civil airline LAPE, emptied of its seats and the hooks on which the sacks of mail were usually hung, has been ordered to bomb an insurgent column between Badajoz and Talavera, on the route taken by Franco's expeditionary force. In the novel, the aircraft has no bomb bay or bombsights and no openings for machine guns to defend itself against enemy fighters. The bombs are piled up, like logs in Nothomb's image, behind the pilots' seats. Great care has to be taken lest they explode. Malraux, alias Réaux, occupies the second pilot's seat, but the pilot is a Spaniard. Interestingly, the author's describes the latter as sourly pretending not to take any notice of Réaux. Réaux suggests very diplomatically that they should descend to three thousand feet before bombing the column. The Spanish pilot agrees but, in order to demonstrate that he is not taking orders from foreigners, he descends to about 6,500 feet only. This is, after all, his aircraft, which suggests that he is a civil airline pilot. Although he is against Franco, he is humiliated to have to accept the aid of foreigners. He would prefer to be fighting in Morocco as he had when he was a young pilot than against his own people. The enemy column appears and the French crew throw the bombs out by hand. The bombs fall in a long sort of necklace. As the aircraft turns away, the crew look back. The column is moving again. It has hardly been touched. As they fly back to base and land, the Spanish pilot feels that both he and the French crew have failed.[46]

British aeroplanes and pilots

Britain was a good market in which Spanish agents of both sides in the civil war might seek to buy airplanes, so much so that the Waldorf Hotel, in the Aldwych in central London, where many of the deals took place, was actually 'bugged' by the British secret service MI5.[47] Actually, none of the machines bought in the UK were warplanes. By 13 August 1936, seven light aircraft had left for Spain, including a Percival Gull sports and touring monoplane which would be used as a personal plane by senior Russian advisers to the Republican forces. Two Dragon Rapides went to each side, but there was no legal ban on civil aircraft flying to Spain and the British authorities, as they awaited full implementation of the Non-Intervention agreement, allowed aircraft destined for Spain to take off. The best way to avoid being turned back by the French authorities when they arrived at Le Bourget was to remove all surplus weight and replace it with extra fuel so that they could fly non-stop to Spain. A few British pilots, often contracted to fly journalists to Spain, stayed on to fly in insurgent squadrons.[48] One of them, Robert McIntosh, who flew journalists' articles from insurgent territory to Biarritz to escape the military censor, was engaged to fly to Portugal to bring back the badly injured Major Ansaldo, who had been the pilot who was to bring General Sanjurjo to Burgos to lead the insurrection and had crashed (see Chapter 2). Next, McIntosh arranged for British Airways, then a private company, to sell four sixteen-seater Dutch-built Fokker F-XIIs to the insurgents. The aircraft were impounded by the French government when they landed to refuel at Bordeaux. When Anthony Eden, the British Foreign Secretary, promised that the Fokkers would not be sold to Spain, they were released and flew back to the new airport at Gatwick. Again they were sold, this time to a Polish firm in Danzig, which lied about where they were going. Instead of going east, they were seen to be flying south. Two crashed, one landed at Bordeaux and one only reached the grounds of a sanatorium near the northern Spanish town of Vitoria, capital of Navarre, centre of the Traditionalist movement. In September, the French released a Fokker from Bordeaux, while Captain Angel Salas flew the plane which had landed in Vitoria to Gamonal aerodrome 4 miles from their headquarters in Burgos. These machines served the insurgents well throughout the war.[49] A DH-89 Dragon Rapide, like the one which had flown Franco from the Canaries to Tetuán, was also sold to wealthy backers of the insurgents and converted into a rudimentary fighter bomber. Arranging for these aircraft to fly to Spain was difficult because, apart from the risks of flying without radio guidance or proper maps, troops often fired on unidentified aircraft and, when they landed, treated the passengers and pilot roughly. Eventually, on 19 August 1936, an Air Ministry *Notice to Airmen* threatened to confiscate the licence of any pilot who ferried a machine to Spain.[50] By that date, ten British aircraft had been delivered to the insurgents and fourteen to the Republicans

The insurgents were supported by German and Italian military pilots and aircraft, but they were not particularly welcoming to the occasional foreign volunteer, who was in any case hardly necessary for their war effort. One British volunteer, Rupert Belleville, Eton-educated and friendly with the Andalusian sherry magnates of Jerez de la Frontera, with a large number of flying hours in his logbook, spent time with the murderous Falange squads in Andalusia, though most of his Spanish adventures were concerned with flying journalists around.[51]

A number of British airmen offered their services to the Republic. Some were commercial pilots. All, save the youngest, had seen service in the First World War. Some had flown in the Chaco War of 1934–5 between Bolivia and Paraguay and others had flown for the Emperor of Abyssinia against the Italians in 1935–6. One of the latter was Hugh Oloff de Wet, a qualified RAF pilot who refers to the inhabitants of Madrid as 'crowds of uncouth people drunk with their new-found freedom, arrogant in their proletarian pomposity', and, more pointedly, 'the plastered proletarian posters of a policy for which I have no sympathy'.[52] Such comments from the son of an officer of the Royal Navy would seem to explain why he offered his services to the insurgents, who declined them. His explanation for flying for the Republic was that, having flown against the Italians in Abyssinia, which was probably why the insurgents rejected him, he wanted to fight them again. Oloff de Wet, who left for Spain in October 1936, was paid £140 per month, plus his hotel bills, an insurance policy, a large bonus for every aircraft he downed and £20 for every pilot he recruited. This high pay would indicate that the Republic saw itself as markedly short of pilots, although the amounts were much smaller than the pilots of Malraux's squadron received. His flying was tested at Heston, predecessor of the modern Heathrow, by another volunteer, Vincent Doherty, also a South African ex-RAF pilot, who had been wounded in Spain. Proving satisfactory, Oloff de Wet was sent to Toulouse to see if he could pick up a plane to add to the diminishing Spanish stock. None was available so he flew Air France to Madrid, which by then was proving risky. Indeed, it is astonishing that a civil air line was still flying into the Spanish capital. It was a very dangerous journey. One of the British mercenaries, Robert Pickett, once mistook an Air France airliner on the Toulouse-Tangiers-Casablanca route for an Italian bomber and nearly shot it down. With bullet holes in its fuselage but not fortunately in its fuel tanks or its passengers, it made an emergency landing at Valencia.[53]

As he landed at Barajas (Madrid) on 4 November 1936, while the insurgent forces were preparing for what they assumed would be their successful assault on the capital, Oloff de Wet noticed broken windows and roofless hangars, the result of daily bombing. Unsurprisingly, the Air France airliner took off as soon as it could, while Oloff de Wet went straight to the Air Ministry, and then and there signed his contract with Colonel Camacho, after which he took a room at the Hotel Florida on the south side of the Plaza de Callao in central Madrid, where many of the famous journalists, including Ernest Hemingway, put up and which is now occupied by El Corte Inglés, a major department store.

The mercenaries were tested at the naval air force base at San Javier in south-east Spain, commanded by the same Major Aboal who had flown to Paris on 21 July with gold coins to pay for French arms. Here, they were given further training in handling the Nieuport-52s and Bréguet XIXs that composed most of the Republic's air force. Doherty, De Wet and two other British mercenaries with them noticed some much more modern-looking aircraft at the other end of the airfield. These were the new Soviet machines which had recently arrived, but De Wet and the others were sent to fly the antiquated Bréguets and Nieuports out of Manises, still the airport of Valencia and, since 7 November, the temporary capital and seat of government of the Spanish Republic. De Wet mentions the corroded instruments and the unreliable compass of his Bréguet XIX and his trigger-happy observer who tried to bomb freighters and destroyers which were almost certainly friendly.[54] When he came across a patrol of German He-51s, he was in a Nieuport escorting Bréguet bombers. Probably obeying wise commands from higher authority, which wished to preserve the Republic's aircraft, he ordered the Bréguets not to confront the thirteen German planes and to turn back to Valencia. The Heinkels then closed with the Nieuports. De Wet had one in his sights but his guns jammed. He felt the cold of petrol on his body because his fuel tank had been holed. If the Heinkel had been armed with incendiary bullets, all his fuel would have exploded and turned his aircraft into a fiery tomb.

Other pilots were sent to Getafe, the base south of Madrid. Until the insurgents overran the base in early November, it was the scene of feverish activity and hurried testing of foreign pilots, which often led to damage to the few aircraft left.

Vincent Doherty, Hilaire du Berrier, an American veteran of the First World War and Abyssinia, several other Americans, De Wet, George Fachiri, an ex-RAF pilot, Robert Pickett, who had stayed in the RAF for only a few weeks, Sydney Holland, Walter Coates, a commercial pilot, and several others served the Spanish Republic in the air. Most of these men were unsettled, both professionally and in their personal lives. A few were heavy drinkers. Many had not been able to tolerate RAF discipline and felt more at ease in the early weeks of the Spanish Civil War where, on the Republican side at least, automatic obedience to authority was out of fashion. None of them seem to have had any particular interest in Spain or any sense of political or ideological mission in flying for the Republic. Some were probably too old and no longer had the reflexes for flying warplanes, especially since these were faster than the ones they had flown in the First World War. Holland, especially, was over 50 years old. Both Holland and Coates had lost their jobs, were family men and were in some financial difficulty. They took the chance of death or serious injury in exchange for the certainty of accumulating a considerable amount of money. They were to be paid £220 per month in sterling, a living allowance of £24 per month, which would be generous for Spain, a bonus of £300 for every enemy machine shot down and £1,000 compensation for them in case of permanent injury and £1,500 for their dependants in case of death. The later Spanish ace and fighter group commander, Andrés García Lacalle, recalled them from his exile in Mexico long after the civil war. Sadly, Sydney Holland was killed on 11 December 1936 when his Monospar ST-25 was jumped by a Heinkel fighter. When Walter Coates got back to England, he made sure that Spain honoured its contract and paid Holland's wife and family the money due to them.[55]

García Lacalle remembered a pilot named Cartwright, who was youngish and elegantly dressed, wearing yellow gloves. The first thing he did when sitting in his fighter was to align the guns, which showed that he had been in military service. García Lacalle, as he wrote his memoirs years afterwards in his Mexican exile, remembered names rather than faces from the desperate confusion of those days when it seemed likely that the insurgents would take the Spanish capital. He recalled Cartwright, De Wet, Doherty, Clifford, Collins and Smith-Pigott. Cartwright, whose real name was Brian Griffin, was the first British pilot to be killed. He died from Italian machine-gun bullets in his back on 27 August at the age of 19 and, as García Lacalle recalled, despite his youth he had spent several months at an RAF training school. He was buried with military honours. However, in his Nieuport he was no match for the *Chirris* flown by the experienced pilots of Dequal's Italian squadron. Clifford's real name was Claude Warsow, also an ex-RAF pilot. He was killed on 25 September 1936. Smith-Pigott's real name was Downes-Martin. He was a wealthy marine engineer from Christchurch in Hampshire, who had a civil pilot's licence and went to Spain for the thrill. He flew one of the Loire fighters and complained that he had to land it at 100 mph.[56] He was killed in the same dogfight on 25 September when an explosive bullet hit him in the back of the neck.[57]Other pilots managed to land their aircraft safely even when they had suffered serious and agonising wounds from the explosive bullets fired from an Italian CR-32 *Chirri* fighter. Doherty, for instance, was hit in two places, but his flying skill allowed him to climb away fast in the rapid-climb Dewoitine that he was, fortunately, piloting.[58]

Another man, who appears to have been slightly unbalanced due to his fracturing his skull in his adolescence, was Charles Kenneth Upjohn-Carter, often known as Charles Kennett.[59] Carter had flown briefly in the First World War. He claimed to have flown for the air mail in the United States and for American gangsters, as well as for civil airlines in other countries, though little can be substantiated except for his frequent court appearances and spells in prison for financial dishonesty. While on bail in London, he offered his services at the Spanish embassy, organized a group of flyers and went to Spain to fly, so he claimed, one of the three Hawker Furies that had remained in Republican hands. Arriving in Madrid on 2 September 1936, they signed the usual highly paid contracts and checked in at the Hotel Florida where the Malraux pilots were noisily celebrating. Next day they were driven to Getafe base, kitted out with Spanish uniforms and shown the aircraft they were to fly: Bréguet XIXs and Nieuport-52s, not the Hawker Fury that Kennett had imagined himself piloting. The squadron they were to join was a mixed Bréguet-Nieuport unit led by Major Antonio Martín-Lunas Lersundi, which included a number of Soviet Russian pilots.[60] There is some question as to whether any Russians actually flew in the international squadron. García Lacalle denies that they did,[61] but Paul Whelan's name-by-name and fully referenced list states that, for example, Yevgeni Erlykin actually commanded the squadron, and that Ivan Kopets, Anton Kovalevskii, Ernst Schach, Georgii Tupikov and Ivan Proskurov flew in it.[62] This is confirmed by the article by General G. Prokofiev in *Bajo la bandera de la España Republicana*, a set of essays by Soviet participants in the Spanish war. Prokofiev recalls that he was shocked by the lack of security. At Getafe, there was no control on who came in and out, and the aircraft were lined up outside the hangars, a prey to any incursion of the enemy.[63]

As well as French and British volunteer pilots, a number of Americans found their way to Republican Spain. They had served in the First World War and since then had flown for Mexican revolutionaries and had barnstormed the length and breadth of the United States. Three of them, Freddie Lord, Gordon Berry and Bert Acosta, were sent to Bilbao to fly a group of unarmed civil planes. Bombs had to be thrown out by hand. There were no gun mounts; the observer of two-man machines would cradle a machine gun and fire it out of the window. Lord once went up in an ill-maintained Bréguet XIX. At two thousand feet, the upper wing collapsed. Rather than risk using the parachutes which he had seen lying neglected on the hangar floor, Lord took a chance and succeeded in landing his plane. With him was his Spanish senior officer, who at once ordered his arrest, but such was the indiscipline on the ground that the Spanish mechanics rescued the American.[64] Very soon there were drunken scenes and serious ill-feeling between the Americans and Captain Manuel Cascón, who was the officer commanding the airfield at Sondica, the present civil airport of Bilbao (Cascón would refuse to leave Spain on the defeat of the Republic in 1939, undergo court martial, refuse to ask for clemency and be executed by Franco)[65]. The undisciplined and drunken behaviour of the Americans led to their departure from Spain in December 1936.

Chapter 5
The Russians arrive

A week after the Spanish Prime Minister, José Giral, appealed to Léon Blum in Paris for aid, he sent a message to the Russian ambassador in France (as yet there was no Soviet diplomatic representation in Spain) asking for aid, though in less specific terms.[1] On 9 August 1936, a Soviet embassy counsellor in Paris informed his superiors that the Spanish Republican authorities were desperate for help. The reply came on 23 August when the Soviet Commissar for Foreign Affairs, Maxim Litvinov, gave the official response: no help of a military nature could be given. Spain was too distant; the cost was too high; the Soviet Union or USSR would honour its agreement not to supply arms to either side in Spain.[2]

Until October 1936, the USSR kept aloof from sending any war material to Spain. Soviet policy was to support Popular Front governments by instructing Communist parties to cooperate with them and to defend parliamentary democracies such as France against the real threat of right-wing revolution. Overall, Moscow desperately wanted to persuade the leaders of the Western world (in the absence of the isolationist United States) that Hitler and Mussolini and not Soviet Russia were the real threats to European peace. Thus in Spain, where in any case the Communist Party was of minor importance, Moscow had not even wanted the Socialist firebrand Francisco Largo Caballero to form a government, as he did on 4 September 1936, because his pre-war revolutionary oratory had been to some extent responsible for the fears which had led to the widespread support for the military uprising of 18 July 1936.[3] Spain was dangerous territory for the USSR because the major working-class groupings, especially the anarchist CNT, were fundamentally anti-communist. In addition, in Catalonia considerable power was held by the POUM (*Partido Obrero de Unificación Marxista*), which Stalinist communists accused of being 'Trotskyist', which it was in the sense of advocating permanent revolution, a policy which contradicted the ideas of the Popular Front.

Soviet policy on Spain and the process of decision-making is unclear. It climaxed in sending aircraft and pilots as well as large quantities of other war material and several hundred military advisers to the Spanish Republic as it fought against Franco and his German and Italian backers. Until October 1936, Russian support for Republican Spain had been limited to propaganda, financial aid and, through the Communist International or Komintern, the organization of volunteers for the Republic's new army in 'International Brigades', whose members were not Soviet citizens. The USSR had accepted the French proposal for Non-Intervention in Spain and seems to have hoped that this would be enough to keep Germany and Italy out of the war.[4] When it was accused by Francoist propaganda of supplying fuel to Spanish Republican warships, the Soviet government took the opportunity to declare,

> The Spanish Government has never asked for assistance and we are convinced that they will find in their own country sufficient forces to liquidate this mutiny of Fascist generals acting on orders from foreign countries.[5]

While the general statement was true, in fact, on 22 July 1936, the Soviet government had instructed its Commissar for Foreign Trade to supply oil to Republican Spain under the most advantageous terms.[6]

Nevertheless, as it became growingly evident that the Republic seemed close to collapse in the face of the aid which the German and Italian governments were sending the Francoists, in aircraft, pilots and considerable quantities of other war material, internal pressures inclined the Soviet leadership to play its part. On 6 September 1936, the Soviet leader Stalin sent a message from his vacation at the Black Sea resort of Sochi to his trusted deputy Lazar Kaganovich suggesting selling fifty bombers, accompanied by twenty first-class pilots, twenty thousand rifles, a thousand machine guns and twenty million rounds of ammunition to Mexico for secret resale to Republican Spain.[7] On 29 September, the Soviet leadership decided that it would approve the plan (which had been submitted on 14 September) to send personnel and armaments to Spain. Throughout that month, as he relaxed on holiday at Sochi, Stalin instructed his subordinates about the volume of aid and the dates when it was to be despatched to Spain. On 26 September, for example, Stalin telephoned Marshal Voroshilov, Commissar for Defence, repeating his suggestion that the USSR should send fifty to sixty fast bombers to Spain. On the next day, Voroshilov told Stalin that thirty aircraft were ready.[8] On 7 October, the Soviet representative in London, presumably under instructions, told the assembled ambassadors who constituted the Non-Intervention Committee which met in the Foreign Office, that, unless German and Italy ceased their violations of the agreement not to send arms to Spain, the USSR would also consider itself free from its obligations under that agreement. By then, Russian and Spanish Republican ships were already at sea with the first supplies of military equipment. This would include several hundred aircraft.

Russian aid was heavily camouflaged. Packing cases were labelled 'Vladivostok' as if they were going to the Far East; the sea routes to be used, of about 2,900 miles, were carefully surveyed, and, once out of the Black Sea and the Dardanelles, and in the Greek Islands, ships disguised their silhouettes. If other ships were likely to sail close, official Soviet sources recount the somewhat unlikely story that the crews strolled along the decks in evening dress, as if they were aristocrats aboard a yacht.[9] As they approached the coast, ships extinguished their lights and changed their flags and names. They sailed close to the Algerian coast until they could be met by ships of the Spanish Republican navy and make a final escorted dash of about 70 miles over the Western Mediterranean into port at Cartagena or Alicante. Radio communication with Moscow was by brief previously agreed signals.

The Russian air force

Russia had had a large air force during the First World War, but the revolution and the civil war which followed decimated it because the majority of pilots, engineers and designers supported the tsarist regime. Not till the latter 1920s were the Soviets able to recover their air strength. With the advantage of official support, the production of Russian aircraft surged massively. In the 1930s, Russian pilots pioneered long-distance flights over the North Pole. By 1936, the USSR was producing state-of-the-art fighter machines and bombing aircraft.

The Russian aircraft industry had to buy aero engines from abroad and was backward in the advanced mass production of airframes, but on the other hand it was well advanced in inventive design. Russian designers, among them Andrei Tupolev and Nikolai Polikarpov, produced aircraft with all the advanced features of the time. Fast Russian monoplane bombers and up-to-date fighters were about to appear in the skies over Spain and surprise their German and Italian opponents.

According to one source, the Soviet ship *Rostok* disembarked thirty-three mechanics and fitters at Cartagena as early as 10 September 1936 to be ready to assemble the Soviet aircraft being gathered for shipping.[10] Soviet shipments to Spain began with the arrival of the *Campeche* in Cartagena on 4 October 1936 after a voyage of eight days.[11] The first aircraft to arrive were ten Tupolev SB fast bombers with complete supplies of spare parts, fuel and ammunition. These arrived at Cartagena on 13 October 1936 on the *Stari Bolshevik*. Thirteen I-15s (I for *Istrebitely* 'fighter') arrived soon after and a further twelve were transhipped at sea from the Spanish ship *Lava Mendi* and taken into Cartagena on small vessels.[12] These twenty-five fighters, unlike the French aircraft which had been sent to Republican Spain earlier, arrived with spare parts, fuel and ammunition. A further ten SB bombers arrived on the *KIM* on 19 October and another ten or eleven on 21 October on the *Volgoles*, making a Russian squadron of thirty-one of these modern fast bombing planes. On 1 November, another fifteen I-15 fighters, under the command of Boris Turzhanskii or 'Maranchov', reached Bilbao in the *Andreyev*, destined for the northern front. On 3 November, the *Kursk* reached Cartagena with fifteen Polikarpov I-16 fighters, followed on 4 November by a further sixteen. On 19 November 31, R-5 two-seater biplanes were disembarked and on 30 December the *Darro* brought ten I-15 fighters to Cartagena. A further fifty I-15 fighters arrived in early January 1937 on the Spanish ships *Sac 2* and *Mar Blanco*. A Soviet squadron of thirty-one RZ (*Radzvedchnik* or 'reconnaissance') light bombers came on the *Aldecoa* on 14 February 1937, and a squadron of thirty-one I-16 fighters arrived on 7 May 1937 on the *Cabo Palos*.

Shipments of Soviet aircraft continued to arrive until August 1937 when sixty-two I-16 fighters arrived on the Spanish ship *Cabo San Agustín*. After this date, no aircraft arrived until 1938 when thirty-one I-16 fighters and sixteen Tupolev SB bombers came on the *Bougaroni* and the *Winnipeg*, followed in the summer of 1938 by ninety further I-16 fighters.

The totals of Russian aircraft shipped to Spain throughout the civil war were as follows:

92 SB bombers
131 or 144 I-15 fighters
276 I-16 fighters
31 R-5 low-level bombers
93 R-Z low-level bombers
4 trainers

The total number of Soviet aircraft, without the trainers, sent to Spain was thus 627 or 640. This compares with 632–732 sent by Germany and over 750 by Italy.[13] It does not, however, include 237 I-15s and 10 I-16s listed as built in Spain. However, whether these were complete aircraft or merely air frames without engines and probably machine guns is contested. Nevertheless, it is evident that the supply of Soviet aircraft declined in the latter part of the war (413 of the 627 arrived between October 1936 and June 1937).[14] As the USSR became more preoccupied with the Japanese invasions of Mongolia and China, its interest in Spain declined. Furthermore, the Soviet leadership was deeply concerned with the loyalty of its military commanders and purged about two-thirds of its senior officers, including some who had flown in Spain. And, however much the USSR and its secret police supported the Spanish Republic as it tried to suppress the social upheaval which had taken place in the Republican zone of Spain, it was not able to convince Britain in particular that the Western democracies should forge a strong alliance with the USSR against the German threat. Thus, in broad terms, while Germany and Italy maintained their rates of aid, the USSR to a considerable extent withdrew from Spain from the latter part of 1937.[15]

Raw numbers are, however, not the only criterion. How far were the aircraft suitable for the conditions of air warfare in Spain? How well-trained were the pilots and air crew? Did the USSR send improved models as the war continued? These and other questions have to be investigated.

The first Soviet aircraft were formed into Group 12, composed of eight squadrons: two of Polikarpov I-15 fighters nicknamed *chatos* or 'snubnoses' by the Republican Spaniards and 'Curtiss' by the Francoists, who insisted it had been supplied by the US manufacturer, two of Polikarpov I-16 fighters nicknamed *Moscas* or 'Flies' by the Republicans and *ratas* or 'rats' by the insurgents, two of Tupolev SB bombers nicknamed *Katiuskas* and two of Polikarpov R-5 *Rasantes* or 'skimmers', which were low-level attack and reconnaissance machines.

The I-15 Chato

The I-15 was a very robust single-seat biplane with wooden wings, metal fuselage and fixed landing gear. At its ideal height of 9,000 feet, it could reach a maximum speed of 225 mph. It was armed with four fixed 7.62mm machine guns and could carry up to 100 kilograms of bombs in underwing racks. It was one of the outstanding fighters of its generation, superior in most respects to the He-51, the fighter sent by the Germans to escort the Ju-52 bomber transports. It was slightly slower than the Italian CR-32 *Chirri* on the level and much slower on the dive, but its rate of climb and manoeuvrability was decidedly superior. Whether its four guns were superior in Spanish conditions of very close combat than the heavier two of the *Chirri* is a matter of debate, as are the relative skills of Italian and Soviet pilots in handling these machines. The first flights, of twelve I-15 machines, were assembled at the base and flying school at Alcantarilla (Murcia) and repainted in the colours of the Republic with red wings and red, yellow and violet rudders, under the command of Senior Lieutenant[16] 'Pablo Palancar' (Pavel Rychagov) and flown to a new airfield called 'Campo XX' at Algete, north-west of Madrid. The other I-15 squadron was based at Alcalá de Henares, to the east of Madrid, moving soon afterwards to Campo X at Azuqueca (Guadalajara), further east, when the I-16s took its place at Alcalá de Henares.

Rychagov was 25 years old. The leading Spanish fighter pilot, Andrés García Lacalle, who flew in his squadron, describes him as young, strong-willed and aggressive, the perfect characteristics of a fighter pilot.[17] Like many of the 772 Soviet pilots who flew in Spain, he had enjoyed a fast career after entering flying school in 1931. He volunteered for service in Spain and arrived together with the first Soviet aircraft. He flew eighty missions in Spain. On return, his career was vertiginous. Nevertheless, in the first days of the war with Germany, when the Soviet air force lost very large numbers of its machines, he was arrested and shot, though he would be rehabilitated in 1954. García Lacalle joined the squadron on 5 November 1936, as the main battle for Madrid was beginning. He was accompanied by four of the best Spanish Republican fighter pilots, Roberto Alonso Santamaría, Fernando Roig Villalta, Alfonso Jiménez Brugués and Jesús García Herguido, together with the French pilot Jean Dary, late of the Malraux squadron. Like García Lacalle, most of these pilots had been corporals or sergeants before the war, which indicates that able Spanish non-commissioned officers had been able to qualify as flying crew. Nevertheless, it also shows how few officer pilots were available to the Republic. These men, whatever their skills, had not benefited from the lengthy flying experience and number of hours in the air that the leading Francoist airmen had enjoyed.

Everyone else at the base, close to Madrid, including the mechanics whose astonishing capacity for work amazed García Lacalle, was Russian, and there was no Spanish representative on the command staff, headed by Yakov Smushkevich (alias 'General Douglas') and Petr Pumpur (alias 'Colonel Julio'). Despite García Lacalle's obvious hostility to the staff at the Ministry of the Navy and Air, his statement that communications from the front went back to Colonel 'Julio', from 'Julio' to the 35-year-old Smushkevich, whose headquarters was at Albacete, and only then to the ministry, now evacuated to Valencia, rings true.[18] Supreme command of the Republican Air Force was held by Lieutenant Colonel Ignacio Hidalgo de Cisneros, but what his position was in the real hierarchy of command remains doubtful.

Whenever the I-15 squadrons were sort of pilots, García Lacalle had the task of summoning others from Valencia or Barcelona, apparently in the absence of any formal organization for replacements. He and other Spanish pilots and foreign mercenaries flew in the first Russian squadrons probably because the Soviets assumed that they themselves would serve relatively short periods in Spain and that the Spaniards, beginning with their best pilots, would have to take over their aircraft. This was a good idea. All the same, some more detailed study of Spanish topography, maps and landmarks under the guidance of Spanish airmen might have saved two Russian pilots who mistakenly landed their planes at Segovia, in insurgent territory some 43 miles to the north-west of the capital. One of these pilots may have been Primo G(h)ibelli, an Italian anti-fascist domiciled in the USSR, whose mutilated body was returned by parachute to the Republican zone with a note: 'This present is so that the head of the Red air force should take note of the fate that awaits him and all his Bolsheviks.'[19]

Undoubtedly, the sudden appearance in the skies above the Spanish capital of the I-15 and I-16 Soviet aircraft was sensational. The confidence that the Germans and the Spanish insurgent pilots had in their Junkers and Heinkels was destroyed when two Junkers, together with two *Chirris* and an Italian Ro-37 fighter-bomber-reconnaissance biplane, were downed on the first day that the

Russians flew, and in the days up to 20 November when the frontal assault on Madrid by Franco's forces was halted. The insurgent general's ground forces, now facing better organized, led and disciplined Republican troops on the western approaches to Madrid, along the river Manzanares and in the Casa de Campo, the spacious heath to the west of Madrid, and among the half-finished buildings of the University City, could no longer count on controlling the skies. The two squadrons of I-15s, known colloquially as *Chatos*, meaning 'snub-nosed' because of their stubby outline, seemed to be the masters. Nevertheless, there were no replacements for those which were destroyed. The first big aerial battle of the war occurred on 5 November 1936 when nine Fiat CR-2s, Italian-led but including the three best Spanish insurgent pilots Joaquín García Morato, Angel Salas and Julio Salvador, met about fifteen *Chatos* and some Potez aircraft over the southern districts of the capital. García Morato shot down a *Chato* and then damaged the engine of a Potez, forcing it to land. Salas shot down a *Chato*, which crashed in flames, and 5 kilometres south-east of Barajas, Madrid's civil airport, he scored hits on two more *Chatos*. He, in turn, came under attack but put his aircraft into a steep dive and made good his escape at treetop level. Salvador chased a *Chato* as far as Barajas and attacked two Potez machines without success.

The insurgent bulletin claimed seven fighters and one Potez destroyed and admitted the loss of one Fiat. The government bulletin claimed that five aircraft had been destroyed. Salas noted in his flight logbook for the day:

> Our nine Fiats met about 15 'Curtiss' fighters. I took one by surprise and shot him down, the aircraft falling some five kilometres south-southeast of Barajas and bursting into flames on impact. I then fired at one head on and later fired at another, before being attacked by two. I managed to shake them off by diving vertically.[20]

In reality, only two I-15s had been destroyed. Lieutenant Mitrofanov, of Rychagov's squadron, was the first Russian pilot to be killed in Spain. Although the other I-15 was a write-off, its pilot survived his forced landing on the wide tree-lined avenue of the Paseo de la Castellana in Madrid. Several other I-15s returned to their base at Campo Soto, near Algete, with varying degrees of battle damage. However, the squadron of *Chatos* seriously damaged two German Heinkel fighters on 6 November 1936 as well as two Ju-52s. Both types of German aircraft were obviously no match for the Soviet fighters.

Russian pilots baling out over insurgent territory were imprisoned while their return was negotiated, sometimes as part of a prisoner exchange. However, for Sergei Tarkhov, who parachuted out of his fighter on 13 November in Republican territory, fate played a cruel hand when the mob, armed with firearms, thinking he was a German, took pot shots at him as he came down so that by the time he landed in Madrid's Paseo de Rosales avenue, he was seriously wounded. Taken to the Palace Hotel, which had been converted into a hospital, Tarkhov died of his injuries several days later. In the USSR, he was awarded the title of Hero of the Soviet Union. Soon afterwards, General Miaja, the Republican commander in Madrid, gave orders that no pilot descending by parachute should be shot at. Spanish insurgents and German and Italian air crew could give important information if carefully questioned.

On 16 November, Rychagov was wounded and replaced by Senior Lieutenant Ivan Kopets, known as 'José'. Kopets would fly two hundred sorties before he returned to the USSR on 17 June 1937, after which he would rise to command an entire air district. He took his own life, probably before Stalin would have had him shot, when the Nazis attacked the Soviet Union and half the entire Soviet air fleet was destroyed in one day.[21]

In early 1937 the I-15 *Chatos* were reorganized. García Lacalle was given the first squadron. The second, which had previously been led by Sergei Tarkhov and then Kosakov (as all the authorities call him, but he was probably Konstantin Kolesnikov, who took over from the seriously wounded Tarkhov. He was killed in May 1937),[22] was now led by Roberto Alonso Santamaría, while the third was commanded by Ivan Kopets and included the Americans Albert Baumler and Charlie Koch. The last named was hospitalized and finally repatriated because his stomach problems could not stand Spanish food, or more likely the strain of flying a fast fighter plane.

The I-16 Mosca

This was an all-metal monoplane with an enclosed cockpit and a retractable undercarriage. It was the first of a generation of fighter aircraft which would include the Messerschmitt Bf-109 and the British Hurricane and Spitfire. The *Mosca* ('fly') could reach a speed of or over 280 mph, faster than any existing fighter. It was armed with two wing-mounted machine guns with a very high rate of fire. Highly manoeuvrable, with the fastest rate of dive in the world, it required very skilled handling. The first thirty-one-plane squadron of *Moscas,* all of whose pilots came from the same squadron in the USSR, was formed into two Spanish-size squadrons. The first twelve machines were assembled and based a few miles east of Madrid at Alcalá de Henares. I-16s made their first appearance over the Madrid front on 10 November when they joined I-15s in strafing Moroccan troops and cavalry in the Casa de Campo, the wide heath on the outskirts of western Madrid. For the Francoist troops, now calling themselves *nacionales*, which British newspapers translated as 'Nationalists', who found themselves on the receiving end of these frequent low-level incursions, the Soviet fighters appeared without warning seemingly from within the abandoned buildings of Madrid's frontline, hugging the ground like rats. Perhaps this is why they nicknamed the I-16s *Ratas*.

All through those autumn days of 1936 German Heinkel-51s and Italian CR-32 *Chirris* fought Russian *Chatos* and *Moscas* over the skies of Madrid as Spanish Nationalist and German and Italian fighter pilots strove to protect the Junkers bombers which were targeting the front line along the Manzanares river and the University City as well as significant buildings such as the Telefónica, Madrid's first 'skyscraper' in the Gran Vía, and the War Ministry on the north side of the Plaza de las Cibeles in the middle of the city.

By the end of November, the Francoist assault on Madrid had eased off. The *Chatos* had fulfilled their task. As for the I-16 *Moscas*, the detachment suffered no losses in December. By the end of the year, the newly arrived Russian fighter pilots had achieved fifteen aerial victories in just two months. On 31 December, seven Soviet pilots were decorated for their performances in Spain, some posthumously, with the Gold Star of the Hero of the Soviet Union and the Order of Lenin.

The number of aircraft downed by the Soviet squadrons is debatable, but García Lacalle's figure of thirty-six during the battle of Madrid does not seem exaggerated, given that he was a participant and indeed critical of pilots' tendency to overstate their victories. Nine Russian aircraft were lost in the battle. García Lacalle was given this figure by Petr Pumpur, alias 'Colonel Julio', a Latvian, born in 1900, who went as a senior commander to Spain in October 1936, returning to the USSR in May 1937. He was awarded the Order of Lenin and named as a Hero of the Soviet Union. After seeing action in China and in the Russo-Finnish war, he was promoted to the rank of general but arrested and shot during the Second World War because a German Junkers-52 had, inexplicably, and to Stalin's anger, been able to reach Moscow.[23]

The SB Katiuska

Andrei Tupolev had designed the SB-2 fast bomber, known in Spain as the *Katiuska* after a character in a recent *zarzuela* or comic opera. Had the Soviets known that the *zarzuela* in question was anti-Bolshevik, they might have objected to their bomber being given the name. The arrival of these aircraft took the Nationalists by surprise. Nothing like them had been seen in Spain before. They were two-engine monoplanes with a maximum speed of 300 mph, a cruising speed of 175 mph and a range of 900 miles. Unable to believe that the Russians, whose aircraft industry was thought to be backward, could have designed and manufactured such an advanced machine, the Francoists called the SB a 'Martin' thinking it a copy of an American plane which a Spanish firm had been considering building under licence when the civil war began. They added the word 'Bomber', which became 'Bomberg' in insurgent propaganda to suggest that its origin was the so-called Jewish-Bolshevik conspiracy to overwhelm Catholic Spain. In fact, the *Katiuska* was entirely Russian and the most advanced bomber of its time. After the first raid by *Katiuskas* on Tablada airfield near Seville, it was only when Lieutenant Ragosin, a White Russian officer of the Legion serving with the Spanish air force, identified the writing on bomb fragments as Cyrillic, that his fellow Nationalists accepted that the bombers were Soviet.

Nevertheless, the SB-2 had severe defects, particularly the vulnerability of its fuel tanks to enemy fire. Its bomb load of 500 kilograms was not impressive. Nor were the engines reliable. From a comment by the Soviet pilot and later General Prokofiev, it would seem that the Russian pilots, though they were recruited from active service members of Russian squadrons, were not familiar with the SB *Katiuska*. As he writes,

> The flight crews of the squadron spent a period of training in order to absorb the techniques of flying the new aircraft. I had not flown that bomber either, but as there was little time available, as soon as I learnt (on the ground) about the instruments I considered myself ready to fly it.[24]

Having to learn to fly the difficult SB fast bomber was only one of the problems that the Soviets faced in Spain, a country with which they were totally unfamiliar and where their interpreters were in short supply and most unlikely to have knowledge of the technical terms required to discuss aircraft and flying. One Russian mechanic complained in his report that the local workers were ill-disciplined. Spanish eating hours and leisurely ways annoyed the Russians as much as they did the Germans. There was little security. Anybody could get on to the airfield and Soviet mechanics suspected there was sabotage, though many accidents and breakages may have been due to mere undisciplined sloppiness.

General problems included, according to V. S. Goranov, an instructor at the Tambov Flight School, that fighter pilots flew sorties in response to ground forces' demands for air support but without time to familiarize themselves with the terrain. Mission details were issued five minutes before take-off, rather than in advance, not allowing time for planning and forcing navigators to do calculations while already in the air. Goranov complained in his report that the SBs were not used for what they were intended. They fought in the skies over Madrid when they should have been used to strike the enemy's rear, his aerodromes and railways. There was chaos at take-offs and it was only by luck that there were no mid-air collisions. It was especially dangerous to take off at night without knowing the positions of the other aircraft, which were scattered all over the airfield. There were no orders pertaining to runway procedures or take-off directions. At first, there were no weather reports either.

Regarding living conditions, Goranov added,

> Living conditions were good, but our rest was insufficient and poorly organized. Dinner was served at 23:00 hours, and as it was at first customary to wait for the squadron leader, his lateness always resulted in extra waiting. Thus, much time that could have been used for sleeping was lost. The Russians were paid 1,600 pesetas a month, but had nowhere to spend them as there was nothing to buy. The French were paid 15,000 pesetas a month, according to their contract. Initially they got up to 30,000, but they lived in luxury hotels in Madrid, whereas the Russians were lodged in a poorer class of accommodation.[25]

A full Soviet squadron of thirty-one SB *Katiuskas* arrived in Spain in mid-October 1936. As part of Group 12 under Colonel Zlatotsvietov, with Pavel Kotov as chief of staff, they were distributed in three squadrons in the part of Spain south of Madrid called La Mancha, home of Cervantes's fictional hero Don Quixote. The 1st squadron was at Tomelloso, in the province of Ciudad Real, and the other two at aerodromes near the provincial capital of Albacete. The commander of the 1st squadron was Major Ernst Schacht, a Swiss who had been domiciled in the USSR since 1922 and had been a member of the international squadron with a small group of Russians among men of many other nationalities in September, flying Potez 54 bombers. Schacht would be caught up in Stalin's purges and shot in 1942, accused of spying for the Germans. This squadron was manned mostly by Russians, though many of the machine gunners were Spaniards.

On 28 October, the *Katiuskas* flew their first mission. Ernst Schacht led eight *Katiuskas* in two patrols, one led by Captain Alexandr Linde and the other by a Spanish pilot surnamed Ramos, which flew 250 miles to the target, which was Tablada airfield, the major air base of the insurgents, near Seville. Crossing Sierra Morena, which separates New Castile from Andalusia, at a height of 6,000 feet over cloud, they descended to 3,600 feet to drop their bombs. The aerodrome responded with anti-aircraft fire. According to Prokofiev, who flew on the mission, fighters were soon in the air challenging the Soviet bombers.[26]

The *Katiuskas* then flew off at a speed which made them impossible to catch, though the insurgent ace Angel Salas in his *Chirri* used his fighter's unequalled diving speed to try in vain to attack them. Four *Katiuskas* made it back to Tomelloso, one was destroyed while making a forced landing and the other was damaged.[27] Over the next few days, the *Katiuskas* twice bombed Talavera airfield and managed to damage six *Chirris* on the ground. However, on 2 November a SB *Katiuska* was pounced on by two *Chirris* diving from high above it. A few days later, two *Katiuskas* collided, and on 7 December the Nationalist ace Narciso Bermúdez de Castro shot down another. By the end of 1936, six SB *Katiuska* fast bombers had been lost of the thirty-one which had arrived in October.

Evidently, Nationalist pilots were getting the measure of the *Katiuska*, which was particularly vulnerable because not only was its pilot's seat unprotected and prey to the explosive shells which the *Chirris* fired but also the fuel tanks were prone to catching fire when hit. The insurgent pilots evolved new tactics to deal with the SB fast bomber. Standing patrols at 15,000 feet allowed the *Chirris* to take advantage of their fast diving speed. On 3 January 1937, the top Nationalist ace, Joaquín García Morato, leading a flight of three fighters known as the *Patrulla Azul* or 'Blue Patrol', shot down two *Katiuskas*, as the Soviet aircraft returned to their base from a raid on the railway station at Córdoba. In his memoirs, García Morato wrote that he suspected an impending attack because the *Katiuskas* always arrived over the city at the same time, and so he stationed himself at a height of 15,000 feet. Spotting the *Katiuskas* on their way back to base at a speed of about 220 mph, he dived at the full speed of the *Chirri* and, firing, saw black smoke coming from one Russian aircraft's engine. He followed it until the pilot, Ananías San Juan, crash-landed it at Andújar, a railway junction some 31 miles from Córdoba. At that moment, García Morato's *Chirri* was hit by machine-gun fire from another *Katiuska*, which the *Chirri* was robust enough to withstand. He turned and, coming up firing under the *Katiuska*, managed to hit it in a vital spot. It fell spiralling and crashed.[28] The two SBs belonged to the 3rd squadron, with the first one being piloted by Spaniard Ananías Sanjuan. He was the sole survivor of the crash landing; a Russian and a Bulgarian pilot were killed, as were the Spanish members of the crew. In the second aircraft, Bulgarian pilot Nikolai Batov (*Ivanov*) attempted an emergency landing in the mountains but the airplane crashed and all three of the crew died.

On the afternoon of 2 November, over Talavera de la Reina, two Italian pilots flying CR-32s managed to bring down an SB, which they still claimed was a 'Martin Bomber' in one of the pilots' vivid words:

> The combat experience that I remember best, and the one I'm most proud of, was when I faced one of the newly-arrived Red Air Force aircraft that were causing us great concern. Dubbed the 'Martin bomber', its appearance came as a rather nasty surprise to us. Thanks to its twin engines and retractable undercarriage, the bomber's level speed was some 30 m.p.h. faster than that of our fighters, which meant that interceptions were impossible. [...] We already knew that one of these famous monoplanes had been sighted over Avila, some 60 miles away [...] I thought to myself that if it comes this way I could intercept it [. . .]
>
> We continued to climb, knowing that the higher we went the faster we could dive down on the enemy bomber. I continued to scan the sky by sector out of habit. First sector, empty. Second sector, empty. Third sector, there, a thin line in the distance against the horizon in a yellowish sky – [...] It was 'him'. I wouldn't let him out of my sight, and I continued to gain height. [...]

> According to my reckoning, it would pass us some six miles away. There would be no chance of getting at him if I didn't change course.
>
> Then suddenly the 'Martin bomber' turned almost 90 degrees and set course for Talavera. The monoplane was well below us, and hadn't seen us. I waited, checked my speed and distance and then at the right moment I pushed the nose of my fighter down and dived at the bomber. The slipstream whistled past and the engine roared [...], but I heard nothing. My eyes were fixed on the aeroplane that rapidly grew larger as we closed in at a tremendous speed. [...] With the first burst I could clearly see the incendiary rounds hitting the wing, sending white sparks flying. The right wing caught fire almost immediately, then a tail of flames from the left wing engulfed both the fuselage and the right wing. The 'Martin bomber' began to fall, but my frenetic dive continued. One wing broke off, at which point three men took to their parachutes. I saw three envelopes open, but the speed was too high and they were torn away – the three men fell like dead bodies.[29]

The SB *Katiuska* was indeed fast, but the Nationalist ace Angel Salas shot down five of them. Generally, the SB could take rapid evasive turns to avoid anti-aircraft fire, but once the Germans brought in their 88mm anti-aircraft and anti-tank gun, sometimes described later as the best artillery in Second World War, its extreme accuracy forced the SB pilots to bomb in a way for which the *Katiuska* was not intended. In spring 1937, for example, two *Katiuskas* flown by Leocadio Mendiola and Armando Gracia set off to bomb a factory in Córdoba, but at 15,000 feet the *flak* was so accurate and so concentrated that they could not begin a flat bombing run and had to dive to 1,250 feet, drop the bombs and pull out quickly.[30] Somehow, the Germans had found the secret of being able to calculate with precision the height and speed of the aircraft.[31]It was true that fighters could not catch the SB from behind but had to attack it head-on or from the side. This might be successful for, according to a report by Lieutenant Vassilii Sharov, who had been a senior pilot in Kiev Military District, the frontwards and sideward fields of fire of the SB were poor. The bomber had many other drawbacks: the pilot's field of view was obscured by the engines, wings and fuselage. The pilot could look forward and upward but he could not see the ground. The SB's engines were of poor quality and the cylinder walls were too thin. One Soviet mechanic reported that ground crew had to replace six out of nine recently installed engines before the planes even took to the air. Joints cracked. Fuel leaked and many other faults of design and manufacture soon became evident: the compass was placed in such a way that the pilot could not use it in flight; the pilot's seat was not fitted with a bomb release lever which he could use in case the navigator was killed.

The Soviet mechanic continued,

> On almost all SBs the upper and the front stabilizer braces broke off in flight. Brackets and bafflers had to be tied down by wire. The celluloid on the F-1 [the navigator's canopy] cracked due to the strong air stream. This also happened when enemy bombs or anti-aircraft shells exploded near the airplane [. . .].
>
> In short, the engine needs improving, as it quickly develops problems.

Perhaps sensing that he ought to be positive, the mechanic added,

> Our designers are correct in trying to develop high-altitude SBs, flying at an altitude of 12–15,000 feet. No fighter is capable of intercepting the SB at such altitude. On the whole the SB is a good airplane and it will justify itself if it is used to do what it has been designed for.[32]

Indeed, on 11 November, *Katiuskas* bombed the airfield near the walled city of Avila, to the west of Madrid, and destroyed several German aircraft. However, on 12 November, two SBs flown by Soviet pilots collided in mid-air due to poor visibility conditions.

Furthermore, the lack of good-quality bombsights and of maps meant that bombing in Spain required the bomber to fly quite low, at which point not only was the SB vulnerable to anti-aircraft fire but also it could be tackled by the new Messerschmitt Bf-109 fighters when they arrived in Spain.

A German pilot recorded in his diary this slightly contradictory comment.

> On 13 November 1936, we encountered the *Ratas* [the insurgents' name for the I-16 *Moscas*) for the first time, and a wild melee resulted. We downed five of them, but what were these victories when compared with the loss of our squadron leader? This only served to show that our good old He-51s were too slow compared with the new *Ratas* – they could play with us as they wanted. Furthermore, the Soviet 'Martin Bombers', which were arriving daily, were 30 m.p.h. faster than us, and the people (i.e. our flyers) were scared of them. Feverishly, we waited for the Bf-109s to arrive from Germany.[33]

Aircraft often crash-landed. Even in friendly Republican territory this could be dangerous for the Russians. A reconnaissance flight over the railway lines in Andalusia and Extremadura by a *Rasante* piloted by Captain Georgii Tupikov, with Prokofiev as observer and another Russian as radio operator and gunner, ended when one engine failed and the other overheated. They landed in an emergency airstrip near Ciudad Real. As the aircraft made a belly landing and gradually came to a stop just short of a precipice, a crowd of angry people stormed over the field towards it, convinced that it must be a German or Italian machine. The Russians refused to leave the aircraft as the crowd demanded, and the gunner fired a burst which forced them to retire, giving time for Prokofiev and Tupikov somehow to insist that the local mayor was summoned. After a stand-off of ten minutes, a man appeared accompanied by rifle-toting militia. Prokofiev could perhaps speak some Spanish, because he spoke to them and said they were from Albacete aerodrome (the existence of the airfield at Tomelloso was secret). The crew had not been allowed to carry any documents of identity in case they were shot down in enemy territory. Fortunately, the mayor (if it was he who had arrived) had enough authority to allow the Russians to remove the guns, the sights and the map. The Russians must somehow have persuaded their captors that they were friendly airmen, because that evening they were driven towards Albacete, a route of about 125 miles, mostly on second-class roads even today, but fortunately passing through Tomelloso. Prokofiev was nervous, thinking that they could easily end up in enemy territory. Gradually the atmosphere became less tense, especially when they stopped in a village and had a beer in a bar. When the Russians took out money to pay they found, as so many foreigners find still today in Spain, that everything had already been paid for and the Spaniards were grinning at them. Finally, they arrived at Tomelloso and, going straight to their hotel, were welcomed by the other Russians of the squadron. It was the turn of the militiamen who had escorted the Russians to be nervous, but as Prokofiev recalled somewhat patronizingly, the Russians congratulated them on their vigilance and from then onwards, fresh fruit, vegetables and meat were sent regularly to Tomelloso. Tupikov himself was shot down later, captured by the Nationalists and repatriated in June 1937. He had a successful Second World War, became a general in command of an air force division and died in 1961.[34]

Less fortunate was the Italian pilot, Lieutenant Ernesto Monico. Shot down on 30 August 1936 in his CR-32 by a Dewoitine fighter on the Central Front between Talavera and Oropesa south-west of Madrid, he demanded to be taken to the Italian ambassador. The enraged militia executed him and claimed, perhaps correctly that they had shot him while he was trying to escape. Monico was the first Italian to die for the Francoist cause.

Much more fortunate on that same 30 August was Hannes Trautloft, the pilot of a German He-51 who bailed out inside Francoist lines. He was able to show his captors a document which read:

Este aparato y su piloto, don Hannes Trautloft,están al servicio del Ejército Nacional del Norte. ('This machine and its pilot, Don Hannes Trautloft, are in the service of the Spanish National Army of the North.')

This, interestingly, indicates that German and Italian pilots did carry identity documents, while presumably the Russians did not.

By March 1937, only nineteen *Katiuskas* were left. The third squadron was disbanded, leaving the other two under the command of Majors Schacht and

Proskurov. Both squadrons took part in the battle of Guadalajara in mid-March 1937, when the Italian *Corpo di Truppe Volontarie* or CTV, the army corps sent by Mussolini to Franco, was routed.

R-5 Rasante

In November 1936, the USSR sent a squadron of thirty-one R-5 *Rasante* or 'skimming' two-seater general purpose (reconnaissance and bombing) biplanes. They became *Grupo* (wing) 15 of the Republican air force, with 109 men including crews and ground staff, commanded by Major 'Vochev'. Soon after arriving, a formation of eighteen R-5s attacked Talavera-Velada aerodrome and destroyed one Italian S-81 bomber and damaged two more on the ground. It was not a successful attack, for four Russian planes were lost. One of the R-5s crashed, hit by splinters from its own bombs. Three others managed to crash-land in the Republican zone, having been attacked by an Italian pilot. He chased the R-5s as they attempted to reach Republican territory, and he succeeded in shooting all three aircraft down. On 4 December, more R-5s attacked Torrijos aerodrome, near Toledo. Again, they were intercepted by patrolling CR-32s and two of them were shot down.

Two days later, they attacked the aerodrome at Navalmoral and damaged six Ju-52s. However, the R-5s were slow and soon withdrawn from action because they needed fighter protection which could not be spared. The R-5 *Rasantes* were reformed into night bomber squadrons.

Possibly a second thirty-one-plane shipment of R-5s in the version called *Shturmovik* [ground attack] arrived in February 1937. This version was armed with four downward-facing machine guns which made it particularly suited for low-level attack on ground troops.

R-Z Natacha

The R-Z *Natacha* was a redesigned, lighter and smaller version of the Polikarpov R-5 *Rasante*, intended for ground attack rather than for reconnaissance missions, but it had a more powerful engine. The R-Z, the first squadron (thirty-one machines) of which arrived in January 1937, saw extensive service on the Republican side of the Spanish Civil War, flying in tight formations and using coordinated defensive fire against fighter attack, while strafing ground troops at low levels. The classic tactic was to enter enemy territory beyond the target, make a wide circle in order to approach the target from behind, then attack in a shallow dive, launching all the 400-kilogram bombload in a single salvo, and shooting off fast towards friendly lines, thus reducing the exposure time to anti-aircraft fire. The observer fired a Shkas machine gun at an extraordinarily high rate of 1,800 rounds per minute, which did not give much firing time but offered a powerful defence to a tight formation of *Natachas*.

Thirty-one *Natachas* arrived in January 1937 and became Group 20, under the command of Major Abelardo Moreno Miró, who figured in the Army List of 1936 as a riding instructor but had at some time transferred into the Spanish air force and achieved his pilot's and observer's diplomas. He received several weeks of training from Russian instructors in low-altitude formation flying. The squadrons of *Natachas* were ideal for low-level strafing of the Italian corps in the battle of Guadalajara. Altogether, three Russian squadrons, a total of 93 aircraft, were shipped to Spain from the USSR in 1937, which seems to suggest that the use of such aircraft for low-level attack on troops was well-understood, and not only by the Germans.

However, when the American mercenary pilot Frank Tinker first saw twenty-four of them in mid-March 1937, he calls them 'huge crates' which flew so slow that the escorting *Chatos* had to fly at close to their stalling speed. However, with a fighter escort and when not exposed to enemy anti-aircraft fire, the *Rasantes* and *Natachas* served as effective ground-attack planes. Frank Tinker refers to their angled machine guns, fixed under their lower wings, which indicates that they were the *Sturmovik* version which, together with the fragmentation bombs which they carried, destroyed a mile and a half of enemy trenches.[35]

The Lacalle squadron

Despite the arrival of Soviet pilots, the Spanish Republic was still in need of foreign volunteers and mercenaries to fly its older aircraft. Some of them would in time fly Russian planes. Among them was 27-year-old Frank Tinker.[36] Tinker had passed flying school for the US Army Air Corps and for the US navy. His tendency to brawling with civilians together with his indiscipline and drunkenness led him to resign his commission in summer 1935 without waiting for the court martial which awaited him. He knocked around for a while until he signed a flying contract with the Spanish embassy in Mexico, a country which was highly sympathetic to the Spanish Republic and would take twenty thousand Spanish refugees from Franco at the end of the civil war. The contract was generous. Tinker would receive US $1,500 each month, and $1,000 for each aircraft he shot down, as well as an insurance policy for $2,000 in the event of his death in action. He was given a Spanish passport in the name of Francisco Gómez Trejo. Not aware that 'Trejo' was the maternal surname and that if asked he should say his name was 'Gómez Trejo', when Tinker said his name was 'Francisco Trejo' and that although his parents were from Spain he had been brought up in Arkansas and could not speak Spanish, the Spanish frontier guards held him for some time, suspecting his bona fides. Nor was any particular welcome extended to him once he got into Spain, for it took Tinker a number of days to travel by very slow trains from the frontier to Valencia and then to the training aerodrome at Los Alcázares in the province of Murcia. After demonstrating his considerable flying skill, he was sent to Manises, today's civil airport at Valencia. Here, in a squadron commanded by an Austrian named 'Katz', he met other mercenaries, among them some ex-RAF pilots. One of them once had the alarming experience of being surrounded in the street by suspicious, gun-toting militiamen, who looked askance at him because he wore a trilby hat, the uniform of the bourgeoisie, at a time when tie and hat had been replaced by the ubiquitous militia overalls, and well-polished shoes had been abandoned for Spanish *alpargatas* or rope-soled sandals.[37]

After a week or two flying Bréguets, Tinker and some other American pilots were transferred. Among them was 27-year-old Harold Dahl, known as 'Whitey' because of his very light colouring and hair. Dahl was later shot down and captured. His wife Edith Rogers, a beautiful singer and showgirl, wrote to Franco pleading for his life, though, with some exceptions where it was perhaps thought that the governments of prisoners were unconcerned, the insurgents did not usually execute foreign citizens. This story later became the basis of the 1940 film *Arise, My Love.* Another pilot was the half-Japanese 'Chang' Selles, whose dismissal was demanded by the suspicious Russians, and another was Albert Baumler, who later flew with the famous 'Flying Tigers' in Burma in the Second World War. While Tinker does not appear to have had political interests, other Americans, among them Eugene Frick and Ben Leider, had gone to Spain specifically to defend the democratic Republic against fascism.

These pilots were sent to another airstrip set up near Los Alcázares to house a squadron of Soviet I-15 *Chatos*. Here Tinker met his new squadron leader, the redoubtable Andrés García Lacalle, the Spanish sergeant pilot who would in due course become commander of the entire fighter force of the Spanish Republican air force. The previous day Tinker had flown a Nieuport fighter for practice. On landing, after fifty yards or so it jumped up in the air, then bounced along first on one wheel, then on another, 'like a frisky pony', in Tinker's words.[38] Tinker came to respect García Lacalle when he learned that, flying Nieuports, the Spaniard had downed eleven enemy aircraft. Tinker, an experienced pilot in the US armed services, was in a good position to evaluate the *Chato*. It flew very similarly to the US navy F-4B, but it had more power. Tinker was impressed by its armament. With four Nadashkeyevich PV-1 7.62 machine guns synchronized to fire through the propeller at 750 rounds per minute, its concentrated firepower was devastating. Like the Italian *Chirri*, the ammunition belts were loaded sequentially with explosive, incendiary and tracer bullets. The pilot's seat was protected by 9mm of armour. The landing gear was built for rough handling, which was fortunate for the bumpy ground of temporary Spanish airstrips and pilots who had relatively little flying experience, for some had only fifty hours, which contrasted with the insistence in the US Flying School for at least three times this in trainer aircraft.[39] Nevertheless, while the Americans pilots in Spain were excellent at keeping formation, they were not very good at aiming their guns either at a sleeve trailed by another plane or at sheets laid out on the ground which they had to strafe after a steep dive and then pull out.

The battle of the Jarama

After the Nationalists had failed to take Madrid from the north and the west, on 6 February 1937 they launched an attack over the Jarama River to try to cut the highway from Madrid to Valencia. Squadrons of *Chatos* under the Russian Kosakov, and others under Roberto Alonso Santamaría, joined García Lacalle's squadron, which was posted to Guadalajara, 34 miles east of Madrid. They were billeted luxuriously amid furnishings 'liberated' from a local mansion. Here also was a squadron of I-16 *Moscas*, whose usual task was to fly high above the *Chatos* and intervene to protect them when necessary. However, only Russian pilots were allowed to get their hands on these sleek and deadly monoplanes. Soon, García Lacalle's squadron was moved to Campo X near Azuqueca. On their first night, while the pilots were asleep in a house in Azuqueca itself, known as *La casa de los pilotos*, the newly occupied airfield was bombed. Wisely, García Lacalle had ordered the aircraft to be widely scattered, and only one had been badly hit.

The Nationalist offensive, which was supported by a battery of the Condor Legion's devastating 88mm guns, which could be used as land artillery as well as in an anti-aircraft role, coincided with an attack from Republican forces in the other direction. García Lacalle's squadron of *Chatos*, with inexperienced Spanish flyers and the American pilots, flew their first mission on 6 February when they machine-gunned a chemicals and gunpowder factory. They downed several He-51 fighters. But losses soon came. The Texan Jim Allison was hit by a shot from the ground, followed by the Spanish pilot José Calderón, hit by flak on 11 February. Calderón, a mechanic, had fled Puerto Pollensa, the seaplane base on Majorca, when it was captured by the insurgents. After the first sortie, García Lacalle gave his pilots a severe lecture. They were undisciplined. They did not keep tightly to their defensive formations. The He-51 and the Fiat CR-32 *Chirris* were very fast divers so, to get away from them, the *Chato* should climb rather than dive. The *Chato*'s turning circle was smaller, so climbing turns were the best manoeuvre.

During an aerial battle on 13 February, an I-16 *Mosca* shot down the commander of the 3rd squadron of Italian Fiat CR-32s, Captain Luigi Lodi, who became a prisoner. Flying his first operational mission, Lodi was at the controls of a CR-32 bis four-gun fighter. The weight of the extra weapons and their ammunition made the CR-32 slower and weakened its wings. Following his loss, all CR-32 bis fighters in Spain had their wing armament removed.

On 18 February, Ben Leider, the only American pilot who had had no flying experience before coming to Spain, lost his life. He had not followed the orders of García Lacalle. The squadron had found itself beneath a large formation of Heinkel fighters. The *Chatos* were heavily outnumbered, so García Lacalle ordered a manoeuvre which the Americans knew and which had been developed in the First World War. García Lacalle signalled his squadron to form a defensive horizontal circle facing left. He had trained his pilots to stay in formation under all circumstances until he brought them out of it. The Heinkels could not penetrate the circle. Ben Leider, the newest of the American pilots, could not resist, left the circle and dived down on a Heinkel, only to be attacked by three of the German aircraft. As he dived, he tried to land in a field but crashed. He was the first American to lose his life fighting for the Spanish Republic, a cause in which the communist Leider sincerely believed.

Later that day, Andrés García Lacalle lectured the surviving pilots. His words were noted by Tinker, who recognized the Spaniard's extraordinary combat flying ability.

> Comrades, today our squadron lost three planes. It was entirely unnecessary. The pilots disregarded the instructions and warnings which had been given them. You had all been told that the Lufberry formation is the best defence in such a situation. And more, you had been told that the Heinkels could dive fifty per cent faster than our ships (*sic* in Tinker's transcription), and you were reminded to think of other planes of the enemy besides the plane at which you were firing. The pilots who are missing forgot all three of these warnings. They left the formation, they knew there were Heinkels above, and they had eyes only for the plane at which they were firing.[40]

So, after a brief lecture on modern aerial warfare tactics, he distributed bottles of beer and sent the pilots back to their quarters. Soon afterwards, the squadron heard with relief that Dahl and Allison had survived their crashes, but with sadness that Ben Leider had not. And on 20 February another Spanish pilot in the squadron, Luis Bercial, originally an aircraft mechanic, was killed in a battle with *Chirris*.

In the air battles of February 1937, in very general terms, *Chatos* flying low in order to bomb enemy trenches and tackle Junkers bombers were protected from Italian *Chirris* by Soviet I-16 *Moscas* which could outdive the Italian fighters. The pilots of the Soviet planes, some Russian and some Spanish, as well as the Americans, completed sometimes several missions a day as they protected the Republican positions from enemy attack from the air and strafed Francoist artillery positions. However, only eight of the original twelve pilots of the squadron were still able to fly. Some advance had been made by the Nationalists. They had cut the railway between Aranjuez and the capital, but the Valencia-Madrid highway was still open. The Lacalle squadron now returned to the Hotel Florida in Madrid for a rest. Here, in March, among sundry other newspaper journalists and war tourists, they met Ernest Hemingway, whose contract with the North American Newspaper Alliance was earning him $500 for each cable – at least ten times more than ordinary reporters for the news service – and $1,000 for his longer mailed despatches.[41]

The spring of 1937 would pit the Soviet aircraft once more, this time against an Italian motorized corps which would try to take Madrid from the north-east.

Guadalajara

The Italian expeditionary force, known as the *Corpo di truppe volontarie* or *CTV*, had enjoyed a walkover when it occupied the ill-defended city of Málaga on 8 February 1937. Now, one month later, under rather different conditions and with a more organized enemy, the corps, consisting of over forty-five thousand men, with artillery, tanks and armoured cars, planned to move its troops on lorries from Sigüenza, north-east of Madrid, along the main Barcelona-Madrid highway towards Guadalajara. The plan was to occupy the latter provincial capital on the first day, Alcalá de Henares on the next and two days later to burst into Madrid. It turned out to be a dismal failure, with huge losses of men and equipment, among other reasons because the Italians did not consider the question of air cover or anti-aircraft protection for their advance. The *CTV* was a very large army corps, commanded by six generals, twenty colonels and a large number of other field and staff officers.[42] Mussolini's ambition was to demonstrate that his new form of fighting, known as *guerra celere* or 'rapid war', which was a poor understanding of the new German theories of mechanized and armoured warfare, would show the 'backward' Spanish military insurgents how a Fascist army could fight. However, the Italian mechanized vehicles were wheeled rather than tracked and in the heavy rain had to stay on the road or risk getting bogged down in the mud. Land vehicles are vulnerable to attack from the air, but the *CTV* advance was not defended by anti-aircraft weapons, the feature which most struck the commander of one of the squadrons of *Chatos*.[43]

In early March 1937, the weather was particularly cold and rainy. The Italians were unsuitably clothed. Many of them were scantily trained Fascist militia, others debt-ridden peasants enlisted for pay, some even claiming on capture that they thought they had been contracted to make a film called *Scipio Africanus* on the deserted high plateau of Castile. Potential support for the advance towards Guadalajara by the *Aviazione Legionaria*, now no longer feigning to be the air arm of the Spanish Legion, whose units were based at Soria and other northern aerodromes, was hampered by bad weather and low cloud. Heavy rain had turned their airfields into quagmires. The Italian squadrons, which should have patrolled over their troops, were unable to take off.

In contrast, the Republican aerodromes either enjoyed hard runways or were sufficiently well-drained to allow for take-offs once the wet weather moderated somewhat. García Lacalle recalls that the airfield at Azuqueca was sown with alfalfa grass which formed a thick carpet with no mud.[44]

In their first attacks with tanks and armoured cars, the Italians overwhelmed thinly held and rudimentarily constructed Republican positions. However, on 10 March, when 'Whitey' Dahl returned from a reconnaissance flight ordered by García Lacalle, his squadron commander, he reported seeing 20 kilometres of the single main road covered with tanks, armoured cars and lorries filled with troops. They were irresistible targets, especially since Italian military traffic discipline was poor and the convoy appeared to be immobilized, but the weather was still too overcast and rainy for strafing. However, on 11 March, there were occasional gaps in the storm clouds. Here was the chance for the Republican forces. The *Katiuska* fast bombers destroyed an Italian division. The *Chatos*, whose role was to spread panic, released their bombs from only 600 feet against troops and vehicles on the road. The Italian lorries and armoured cars were bumper to bumper on the highway while men were furiously striving to push broken-down trucks out of the way. The Italian troops had not been trained to react against strafing, how to disperse or how to train their weapons against low-flying planes. The trenches that they dug soon filled up with water. They bunched around what they hoped was the shelter provided by trucks and armoured cars. There was no defensive fire nor was any cover on the treeless plain, now a quagmire through which the unfortunate Italian troops dragged their mud-laden boots.

Later that evening, the Lacalle squadron was caught in a thunderstorm. Frank Tinker and 'Whitey' Dahl, unable to locate their own airfield, flew to headquarters at Albacete where they were invited to dine with Yakov Smushkevich, the overall Russian commander. Next day, they escorted Smushkevich as he flew his own *Chato* to Alcalá de Henares. On 12 March, a squadron of *Moscas* with Boris Smirnov in command provided cover for the *Chatos*, *Rasantes* and *Natachas* as they destroyed another Italian division. On 14–17 March, about ninety Soviet aircraft supported Republican units which pushed the CTV back. As the month ended and the aerodromes from which the Italian planes were growingly able to take off became usable, the battle ceased. It had been a complete rout for the *CTV*.

Frank Tinker, the American mercenary, writes vividly,

> I spotted one especially large group of Italians in wild retreat before a couple of tanks. My first move was to manoeuvre my plane to a down-wind position

> from them – the wind was blowing in the same direction they were running – and then push its nose over into about a sixty-degree dive. At that altitude – 1,000 feet – the men looked like a mass of ants on the ground, even through my telescopic sight. At about 700 feet I opened fire with one upper and one lower machine gun. This was so I could see, by the tracer bullets, whether or not I was on target. The stream of bullets was just ahead of the fleeing group, so I opened up with the other two guns and pulled the plane's nose up a little.
>
> By this time I could see the individuals plainly; they had also become aware of my presence. Then they did the worst thing they could have done – started running in the opposite direction. I could see dead-white faces swivel around, and, at sight of the plane, comprehension would turn them even whiter. I could see their lips drawing back from their teeth in stark terror. Some of them tried to run at right angles, but it was too late; already they were falling like grain before a reaper. I pushed the rudder back and forth gently, so that the bullets would cover a wider area, then pulled back on the stick just as gently thus lengthening the swathe. I pulled out of the dive about twenty feet off the ground, zoomed up to rejoin the squadron, and started looking for more victims. We kept this up until our gasoline and bullets were so low [each of the twelve *Chatos* was armed with 3,000 rounds] that we were forced to return to the field.

Tinker, thoroughly trained in the United States, understood what he was doing. He referred to it as a 'murderous dive [which] will give you an idea of what the foot soldiers may expect in the war to come'.[45] He describes in detail how he escaped from a fast-diving *Chirri* by allowing the Italian pilot to achieve too high a speed to pull out from his own dive, and making a complete 380-degree turn, coming out behind the *Chirri* and doing the same manoeuvre with the next one. As he writes revealingly,

> This was a good illustration of the advantage that experience gives to a fighter pilot. If I had become involved in a situation like that in my first dogfight I would have headed for the clouds [beneath] in one long dive and would certainly have been shot down. As it was, I knew I could outmanoeuvre the Fiats [...] Experience is still the greatest teacher.[46]

Andrés García Lacalle, commander of the squadron, now suffering from extreme exhaustion, had so impressed his superiors and the Soviet advisers that he was promoted to major, having begun the war as a sergeant, and left the squadron.

Chapter 6
The Condor Legion and the campaign in the north of Spain

All through November 1936 hundreds of men, guns, aircraft and many tons of other equipment were shipped out of Germany's Baltic ports bound for Franco's Spain.

Lieutenant Colonel and Doctor of Engineering Wolfram Baron von Richthofen, the erstwhile head of Testing and Development in the German Air Ministry and future chief of staff of the Condor Legion, flew to Seville via Rome on 29 November. In Spain he took command of *Versuchs-Kommando/88*, an embryonic testing and evaluation staff whose task was to provide formal assessment of the performance of the Legion Condor's aircraft in combat. His first impressions were not good:

> My accommodation is a very bad room in the Hotel Andalucia. From the *Versuchsgruppe* there is as yet nobody, but just some materiel, a part of which is already lying around at the harbour and in an area of Tablada aerodrome. Transport and distribution, information on the arrivals of the steamers and loading lists are all completely unknown. The unloaded materiel is often unusable since many of the important items are often missing. Am greeted by the chief of staff, Holle, who is worn out and wants to be left in peace. I report to General Sander [code name for General Sperrle], who complains about the complete lack of knowledge in Berlin of local conditions here.[1]

Once the Soviet aircraft had gone into action in defence of the Spanish Republic, it was clear that not only would the Junkers bombers and Heinkel fighters of the Condor Legion have to be replaced by more advanced models, but that detailed problems of maintenance and the supply of spare parts would also have to be solved. The fronts on which the Spanish command required the Condor Legion to operate were at such distances and Spain's narrow roads and single-track railways were in such a poor condition, often winding for miles through mountains, that rapid transfer of men and vehicles was often difficult. Accidents became frequent, even when the men were not joyriding – as was a not uncommon practice given the availability and potency of local wine – while tyres wore out and engines overheated. The efficiency of communications was discovered to be insufficient. Telephone and telegraph lines were cut and not infrequently tapped by Republican guerrillas. Rain caused short circuits. The winter weather shut down airfields and paralyzed land operations except in the Málaga area, where a Condor Legion liaison officer closely watched the Italian walkover and occupation of the city on 7/8 February 1937 for the lessons that might be learnt about mobile warfare, in which the Italians were thought to be superior. However, problems turned out to be ideal exercises for Sperrle and von Richthofen and their staffs, who solved the even more delicate issues of liaison with their Spanish Nationalist allies. Much of the detail of what was developed in Spain under the stress of combat became standard procedure for the Luftwaffe.

This was true especially of logistics. Everything save food, and particularly munitions and fuel, had to be shipped from Baltic ports through the North Sea and the English Channel as far as Galicia in north-west Spain. Key personnel and small items arrived throughout the war by four weekly flights direct from Germany to Spain. Only two of these crashed en route, one leading to the death of Major Scheele, the first German commanding officer in Spain, on 24 February 1939 just before the end of the war.[2] Replacements of three or four hundred men, among them pilots, mechanics, armourers, communications specialists and medical personnel, were regularly assembled at an airfield near Berlin, issued with civilian clothes, moved to Hamburg and transferred on barges to a dimly lighted ship moored in the far reaches of the harbour. Security at sea was so tight that no man was allowed on deck if any vessel came within telescope distance.

The voyages in the winter were not as pleasant as the first sailing on the *Usaramo*, back in the summer of 1936. The later Luftwaffe general Adolf Galland recalled that, having passed the rigorous selection procedure to enter the Lufthansa training programme, and having flown the Stuttgart to Barcelona route, he completed a still secret training programme for military pilots and joined the Luftwaffe. In April 1937, having noticed familiar faces in his squadron suddenly disappearing and returning months later with tanned faces, Galland responded to a confidential invitation to volunteer for service in Spain. He reported to *Sonderstab W* where he was sworn to strict secrecy, given the coded address for mail of 'Max Winkler, Berlin SW88' and sent to the base at Döberitz. There, in company with 370 other men disguised, like the previous expedition, as a *Kraft durch Fried* group, he was issued with civilian clothing, taken to Hamburg and embarked on to a ship flying the Panamanian flag and thus not liable to be investigated by any warship belonging to the European signatories to the Non-Intervention agreement. After the Second World War, Galland told the American officer and historian Raymond Proctor that the ship had been previously used to ship Soviet arms to Spain, had been captured by one of Franco's ships and was in a deplorable state.[3] The weather was bad and many men were seasick. It took twelve days to sail to El Ferrol on the coast of Galicia, which they reached on 8 May 1937. When they landed, they were issued with the brown uniforms of the German Labour Service or *Reichsarbeitsdienst*.

The usual procedure was for a Nationalist ship to escort Condor Legion replacement vessels into the harbour of El Ferrol – Franco's birth place – at night out of the sight of one of the British warships which made it their business to observe what was going on in all Spanish ports, thus providing much information for the Admiralty and the Foreign Office. The Germans would be entrained for León, from 1937 onward the Condor Legion's main depot, or to Vitoria in Navarre where many of the aircraft were based until the war in northern Spain ended in October 1937.

Fighter aircraft of the Condor Legion

The arrival of a shipment of sixty crated He-51s, disguised as agricultural machinery, in Seville on 18 November 1936 for assembly at Tablada heralded the adoption of the new unit designation for the fighter group of the Condor Legion. This was *Jagdgruppe 88* or just J/88, led by Major Baier, followed soon by Captain Hubertus Merhart von Bernegg. The three new fighter squadrons were established as 1.J/88 under Captain Werner Palm, 2.J/88 under Captain Siegfried Lehmann and 3.J/88 (intended to undertake ground-attack sorties) under Captain Jürgen Roth, who led a batch of recently arrived pilots. The original cadre of Heinkel pilots already in Spain became the 4th squadron or 4.Jagd/88 under the leadership of Herwig Knüppel, now promoted to captain after the death of Lieutenant Kraft Eberhardt on 13 November. Knüppel recorded:

> I now had to take over the leadership of the Eberhardt squadron. In the period that followed [winter 1936/37] we were pitched into all the battlefronts and thereby got to know the whole of Spain. León, Burgos, Vitoria, San Sebastian, Logroño, Zaragoza, Teruel, Barahona, Ávila, Escalona, Córdoba, Almorox – these were our combat airfields. The entire squadron consisted of only 35 men, but we stuck together like *Pech und Schwefel* [pitch and sulphur].
>
> The ground crews worked untiringly [. . .] Once, in a low-level attack by enemy fighters, the armourer *Unteroffizier* Eick, standing completely unprotected,

> shot down freehandedly one of the attackers with a rifle. This all goes to show how excellent the fighting spirit of our comrades was.[4]

Like their predecessors, the new German pilots viewed operating conditions in Spain with shock and disdain. Harro Harder lamented that at Tablada

> [the] situation was awful. We would welcome an opportunity to sort things out. The fighters sit here and don't go anywhere. The entire operation appears increasingly like some great escapade controlled by incompetent staff officers. Are our operations justified by results? Why can't we have better aircraft?[5]

In his diary, Otto von Winterer complained, in early 1937, about the competence of the German commanders. Regarding von Richthofen, he wrote words which would almost certainly have got him dismissed from the Luftwaffe if his diary had not been captured by the enemy when von Winterer was shot down. The translation is worth quoting when it was probably written in the context of losses against the far more modern Soviet fighters.

> He gets the salary of a major, but he can't even fulfil the role of a foreman.[6]

Similar comments were made about other German commanders.

The first fighter squadron (1.J/88), which comprised eleven pilots, was just about operational by the end of November and was moved north to Burgos. However, by early December, the German fighters were ready for operations from their bases at León and Vitoria, and near Madrid at Ávila and Escalona. Winter in the high parts of northern and central Spain is very harsh. While the high command moved into a castle and other personnel were billeted in private houses, the new men were housed in their *Wohnzug* or 'housing train', which consisted of carriages for officers and men, cooking facilities and a dining car and a steam engine at either end which could move the train rapidly to wherever the men were needed and, with steam constantly up, offered a warm place to sleep and relax after a day in the freezing winds of the airfields. The communications section was based close to Franco's headquarters at Salamanca and fully connected to all Condor Legion bases and Spanish and Italian commands.

The pilots of the 2nd squadron decided to create an identity for themselves, as one remembered:

> At this time, we painted top hats on the aircraft, and we soon became known to the Spaniards as 'la escuadrilla con los sombreros' ('the squadron with the hats').[7]

The new squadrons of He-51s commenced sorties over Madrid in late December 1936. The German fighter pilots evaluated various tactical methods both to attack enemy aircraft and to defend friendly bombers. Nevertheless, given the superiority of the newly arrived Russian fighter aircraft, General Sperrle, commander of the Condor Legion, restricted the fighters' escort duties to brief sorties at dawn and dusk. One freezing day in January 1937, a company commander in the Thaelmann battalion of the International Brigades of the Popular Army of the Spanish Republic watched as a German fighter plunged in flames over the Corunna highway a few miles north-west of Madrid. When the flames died down, he found the corpse of Lieutenant Hans-Peter von Gallera, a former comrade in a Luftwaffe squadron.[8]

On 30 November, von Richthofen attempted 'to compile an overall picture'. He noted in his diary:

> Red air attacks appear to be gradually setting in. In the last 14 days, they have increased from two to three, and they have now made six bombing raids on Seville and Cádiz. These are only frivolous and without any effect. Red fighters have only been seen up to now in the Madrid area. *Our own operations there, without fighter protection, by day, are considered impossible*. At other locations, no Red fighters have been observed. However, if we conduct daylight raids, their surprise appearance is feared as a probability.[9]

At the start of 1937, given the He-51's technical shortcomings and Republican fighter supremacy, the Condor Legion failed to achieve air superiority and suffered an alarming 20 per cent loss rate. Lieutenant Harro Harder commented:

> We were all convinced that it was madness to continue sending the He-51s on escort missions over Madrid. The Soviet I-16s, known as *Ratas* to the Germans and Franco's air force, played cat and mouse with us. Even the 'Martin' bombers [SB *Katiuskas*] were at least 50 km/h faster than us. The morale of our pilots was excellent, but all the guts in the world were useless with such technical inferiority.[10]

In mid-January, Captain Hubertus Merhart von Bernegg, commander of the German fighters of *Jagd/88*, drafted a report to General Sperrle in which he protested at having to tackle Soviet aircraft with the He-51 and advised that he would no longer be sending his men on missions to engage the enemy. This was insubordination and an infuriated Sperrle immediately flew to J/88's headquarters and was met by Merhart as he disembarked from his aircraft. A tense 'face-to-face' showdown ensued in which Merhart defiantly refused to sanction operations against the enemy air force and threatened to ask to be relieved of his command and sent home to Germany. After a moment in which he composed himself, Sperrle turned on his heel and strode back to his aircraft.[11] Nothing more was heard of the matter, until orders were received changing the entire tactical deployment of the He-51. To his credit, and belying his fierce appearance and his ordering the arrest and court martial of Captain Merhart, Sperrle had taken up the matter with *Sonderstab W*, which was well aware of the problem, and from Berlin General Wilberg now directed that with immediate effect the He-51 would fly only 'low-level attacks against enemy frontlines'.[12] In addition, these German fighters, diving down guided by enemy lorry headlights, would attack transport on narrow roads, bridges and in the villages through which the main roads to the capital ran. The new orders show how greatly a better supply of steadily improved Soviet aircraft would have inhibited the Condor Legion.

Nevertheless, in their evaluation of performance in air-ground attack, the Germans, well-used by their training to the careful study of results, noted just how difficult it was to score hits on bridges. Nor was it easy to spot the multiplicity of small airfields where *Chatos* and *Moscas* could be easily hidden and camouflaged. German reconnaissance aircraft had no way of taking pictures at night. The Legion's major contribution was to be ground attack, using the He-51s in a new role against fortified positions so that Nationalist infantry did not go forward into a hail of machine-gun bullets like Allied and German troops in the First World War, sometimes losing over half a battalion in taking one line of trenches. This tactic required close liaison with the ground and, in this case, with Spanish troops. While at times the Republican forces were so ill-prepared that they abandoned their positions after being attacked from the air, at others the German pilots complained that the low-level Nationalist infantry officers did not immediately order their men forward to take advantage of the enemy's disorganization following a strafe by the Condor Legion.[13] This was the beginning of the Condor Legion's steep learning curve which made Spain such a valuable lesson for the Luftwaffe. Although low-level strafing had been used in the First World War, the new Luftwaffe had no experience of it. As Condor Legion pilot Douglas Pitcairn told historian Raymond Proctor after the Second World War, 'We learned the low-level ground support operation in combat as On-the-Job Training.'[14]. A difficult manoeuvre in low flying was to avoid high trees and to swerve away from rifle and machine-gun fire from troops who had the necessary high morale and determination to fire at a fighter as it dived down on them with its machine guns blazing. The German pilots also had to learn how to avoid their own bomb blasts. The Condor Legion pilots also developed hand-to-hand signals for pilots to communicate while in flight, and panels, flares and smoke pots for ground-to-air communications.

The withdrawal of the He-51 as a fighter aircraft and its deployment in the close-support role of battlefield support signified a change in mission. Consequently, Captain Siegfried Lehmann's 2.J/88 squadron with its ten He-51s was moved to Vitoria, in northern Spain, and another squadron was sent to León. From Vitoria, the Heinkels engaged in regular ground-attack missions, dropping fragmentation bombs on enemy positions and strafing road transport. The other units remained on the Madrid Front at Escalona and Ávila.

Plate 1 A historic image: The Junkers 52, on Thursday, 23 August 1936, with its passengers, just before taking off from Tetuán for Germany. Left to right: Captain Arranz, Johannes Bernhardt holding Franco's letter appealing to Hitler for arms, the pilot, unidentified, and Adolf Langenheim.

Plate 2 Moorish troops waiting to board German aircraft.

Plate 3 Guernica in ruins 1937.

Plate 4 Servicing a Messerschmitt 109 fighter.

Plate 5 The Condor Legion about to embark for home.

Plate 6 Hitler and von Richthofen review the Condor Legion after its return.

Plate 7 In the centre, Rómulo Negrín, son of the Spanish Republic's Prime Minister.

Plate 8 Yakov Shmuskevich, commander of Soviet Air Force in Spain. Courtesy of ADAR and the Archivo de la Democracia (Universidad de Alicante).

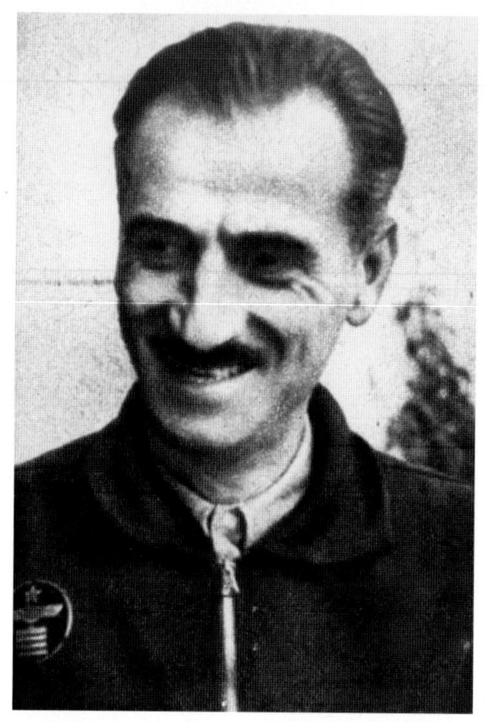

Plate 9 Ignacio Hidalgo de Cisneros, commander of Spanish Republican Air Force. Courtesy of ADAR and the Archivo de la Democracia (Universidad de Alicante).

Plate 10 Manuel Cascón (1895-1939), led expedition of trainee pilots in USSR, surrendered Republican Air Force, executed by Franco. Courtesy of ADAR and the Archivo de la Democracia (Universidad de Alicante).

Plate 11 A Soviet *Chato* over Central Spain.

Plate 12 A French Potez bomber downed in the service of the Republic.

Plate 13 A Soviet SB *Katiuska* bomber.

Plate 14 Mussolini reviews his air force shortly before the Spanish Civil War.

Plate 15 Franco greets Hitler as he descends from his train at Hendaye on 23 October 1940. Hitler hoped in vain that Franco would join the war on his side. Franco's brother-in-law and Foreign Minister, Ramón Serrano Súñer, a strong advocate of Spain's entry into the war, stands at Hitler's right.

It now became essential to send newer aircraft, many of them still in the experimental stage, to Spain. In December, the VJ/88 (*Versuchsjagdstaffel* or Experimental Fighter Squadron) had been despatched to the base at Tablada, near Seville, bringing a small number of Messerschmitt Bf-109 prototypes with different types of propeller and armament. However, the ports and roads were choked, and German reports complained that the Bf-109s were 'lying around' a harbour, probably Cádiz. Three days later, von Richthofen bemoaned that

> transport roads from the harbour to the airfield are so blocked up that the crates with the Bf-109s, locked inside the wagons for three days now, cannot be brought to the airfield. Everyone has now been notified. Any success likely?[15]

On 5 December, Lieutenant Hannes Trautloft and *Unteroffizier* Erwin Kley were selected to test the first Messerschmitt Bf-109s,[16] which were expected to be ready within three or four days for local operations over Seville. Von Richthofen was irritated by the delays and further angered when on 9 December Kley destroyed a Bf-109 in a crash on take-off at Tablada. It took until 17 December for all six experimental Bf-109s, early versions of the fighter which would fight the RAF's Spitfires and Hurricanes in the Second World War, to be unpacked and assembled. These were experimental planes and returned to Germany. However, General Sperrle had been impressed and urgently demanded a squadron of these new fighter monoplanes. Sixteen arrived in March 1937 where they were to be flown by pilots of 2 J/88, the second fighter squadron of the Condor Legion. In the meantime, the Condor Legion turned its some of the He-51 fighters over to Spanish crews, though it retained two squadrons for ground attack.

The Messerschmitt Bf-109

By April 1937, the relatively rapid conversion of the Condor Legion's fighter pilots was complete and some Bf-109s flew to Vitoria from where they could take part in the campaign in northern Spain.

Not until the end of the war did the Russians send a fighter plane to Spain which could rival this famed German machine, which from 1937 onwards would dominate the skies over the Spanish battlefields. The Bf-109 was a single-seat, all-metal low-wing monoplane, with an enclosed cockpit and retractable landing gear. Following the experimental prototypes which had already been sent and withdrawn, the first sixteen production models were sent to Spain in March 1937 to re-equip the 2nd squadron (2J/88) under the command of Lieutenant Gunther Lützow, who supervised the conversion training of his crews from the He-51 biplane, with its open cockpit, its fixed landing gear and its top speed of 205 mph. This first production model of the Bf-109 could fly at a maximum speed of 290 mph at low level, and at its ceiling of 28,000 feet it cruised at 235 mph.[17] This rivalled the Russian I-16 *Mosca* at least up to about 3,000 feet, after which the *Mosca*'s engine became less powerful. The Germans would steadily replace the Bf-109 with models with higher specifications. In 1938, they would be able to fly higher and for longer and they would be equipped with air-to-ground radio. Beginning with two synchronized machine guns mounted in the cowling, firing over the top of the engine, and a third through the propeller arc, their weaponry would be augmented by guns in the wings and eventually by two cannons which fired 20mm bullets, as well as some fragmentation bombs which were small but deadly when they exploded among crowded enemy troops.

In direct competition with the Bf-109, the new Heinkel-112 began its tests in Spain on 9 December 1936. It would be used primarily against Republican armour during the offensive in northern Spain in the spring of 1937, with some effect, earning the fighter, which already bore the name *Kanonvögel* ('Cannon Bird'), the additional nickname of *Dosenöffner* ('Can Opener') for its destruction of three enemy tanks during the advance on Bilbao. However, not till late 1938 were further versions of this aircraft sent to Spain.

German pilots were hastily brought home in order to undergo conversion courses to fly the new machines which now began to arrive in Spain from Germany. After a month, von Richthofen returned to Berlin where on 6 January 1937 he conferred with the leaders of the Luftwaffe. He returned to Spain with orders to take over the post of chief of staff of the Condor Legion.

Bombers of the Condor Legion

A major problem suffered by the Ju-52 bombers on which Franco's German allies had relied since July 1936 was their limited practical ceiling of 11,500 feet. In the first winter of the war, cloud and icing up were major challenges. In some weather conditions, the Junkers could not fly over the series of mountain ranges in Spain, the highest country in Europe save Switzerland. As so often, the Spanish Civil War forced the Germans to face problems to which they were not accustomed. Weather-reporting stations had to be established, and this was a move of considerable value for the future when the weather over distant targets became a matter of great import.

In January 1937, von Moreau's original squadron of Ju-52 bombers was dissolved. Many of the aircraft were transferred to Spanish command. The crews were rotated back to Germany where their combat experience was invaluable in training the pilots of the 4th bomber squadron – in Luftwaffe parlance 4 *Kampf*/88 – which became the *Versuchsbomberstaffel* (experimental bomber squadron) or VB/88, equipped with twelve modern machines: Dornier-17s, Heinkel-111s and Ju-86s. These had been developed from prototypes which, with the use of the most up-to-date structural and aerodynamic advances, could serve as twin-engine all-metal monoplane bombers as well as mail and passenger transport machines for Lufthansa. They were shipped to Spain in early February, assembled in Tablada's huge hangar, built for the Zeppelins of the German passenger and postal service which landed in the Andalusian capital on their way to South America and then flown to Salamanca in western Spain. The Dornier, known as the 'flying pencil' because of its very slim fuselage, had a maximum speed of 220 mph, its range was 410 miles and it carried 500 kilograms of bombs. As successors to Von Moreau's original squadron, the Dorniers were nicknamed 'Pablos'.[18]

The first thirty Heinkel 111 fast bombers began to arrive in Spain in February 1937. These new machines were seen in action over central Spain flying their first mission on 6 March 1937. This was a strike against the Republican bases of Alcalá de Henares and Barajas, east of Madrid. The He-111 had a maximum speed at 13,000 feet of 230 mph and a bomb load of 3.3 tons. On 8 and 9 March 1937, He-111s from Avila flew over the various Republican airfields though not hitting much of consequence,[19] but their speed, at least when the ventral gunner's projecting pod was removed, compared with the lumbering Junkers bombers which had required fighter escorts, meant that Republican fighter pilots would need to be retrained and to readjust their techniques, as the Germans and Francoist pilots had had to do in order to tackle the even faster Russian *Katiuska* bombers. This was taking place in the international context of decision-taking about what was required in a bomber aircraft. Could bombers be built which flew so fast that they did not require fighter escort? If so, it would signify that the famous statement of British Prime Minister Stanley Baldwin that the bomber would always get through was accurate, unless the development of fast bombers also meant that national aircraft policies would have to decide whether or not to concentrate on building such fast fighters that bombers could be effectively challenged.

Campaign in northern Spain

After the Italian disaster of Guadalajara in March 1937, Franco desisted from trying to take Madrid. The city, heavily defended by five Republican army corps, would remain a symbol of the resistance of the Republic until March 1939 when the Republic would collapse through inner disagreements and an officers' uprising against the government.

Franco and his field commander in the north, General Emilio Mola, now turned their attention to the Republican fringe along the northern coast of Spain. The western Pyrenean frontier with France and the fashionable seaside resort of

San Sebastián had been secured by the Francoists in 1936, but the Republic held the stretch between that city and Bilbao, capital of the Basque Country, an autonomous region since October 1936, and from Bilbao westward to Santander, and further westward as far as the mining district of Asturias and the port city of Gijón. However, not only was the northern Republican zone relatively weak and disorganized in military terms but it also was cut off from the rest of the territory of the Republic and politically divided. Furthermore, Franco's navy was aggressively exercising a blockade against merchant ships bringing war material into the northern zone, particularly into the port of Bilbao. The Basque region, even though fiercely defensive of its autonomy, granted in October 1936 but loathed by the centralizing Francoists, was fundamentally conservative and Catholic. The Basques were not greatly in sympathy with the Republic where churches were closed and thousands of clergy had been murdered. There were serious disagreements between José María de Aguirre, the Basque president, and the Republican war leaders, the socialists Francisco Largo Caballero and Indalecio Prieto, now resident in the Spanish Republic's temporary capital of Valencia on the Mediterranean coast. The port of Bilbao would, if taken, enable Franco's army to receive German supplies more quickly than further along the coast at Vigo and La Coruña. Furthermore, the Basque region, with its iron ore and developed steel and shipbuilding industries, would be highly valuable to the Spanish insurgents. Thus, the northern zone would be Franco's main target in the spring and summer of 1937.

From the point of view of the war in the air, Republican territory along the Cantabrian Sea was not only narrow but also mountainous and unsuitable for airfields.[20] Furthermore, aircraft flying from the nearest Republican airfields to Bilbao had to cover many dangerous miles over enemy territory as well as over mountainous regions. While occasional larger and fast transport aircraft such as the DC-2s of LAPE could manage the distances and were too fast for fighters to attack, such flights were chancy for Republican fighters with limited fuel capacity. In contrast, the Nationalists could maintain their air superiority in the north because they could fly with equal ease over both the northern and central Republican zones from Burgos, where there were now eight He-45 and He-70 reconnaissance aircraft and twenty-two bombers of the K/88 and VB/88 experimental squadrons, and from Vitoria, with three squadrons of fighters. These were called 'Marabou' and 'Mickey Mouse', equipped with H-51 biplane fighters, and 'Top Hat', with seven Bf-109s. In the Condor Legion's order of battle, there were also some H-45 light bombers, flown by Spanish crews, and three Henschel-123 dive-bombers.[21] This aircraft was the first dive-bomber (*Sturzkampfflugzeug*, usually known as *Stuka*) to be built in quantity for the Luftwaffe. The question was, could the dive-bomber hit specific and relatively small targets more accurately and from a greater height than well-directed artillery fire? Von Richthofen, among others, doubted their ability to do this. Later in the

Spanish war, the failure to hit bridges with horizontally flying bombers would lead to a change of mind and the increased use of the dive-bomber. This would be yet another example of how greatly German participation in the Spanish Civil War perfected the Luftwaffe's techniques in the Second World War.

On 29 March 1937, the Condor Legion moved its staff to Vitoria, in Navarre, ready for the infantry offensive which was to begin against the neighbouring Basque Region on 31 March. In the Nationalist ground attack, the Francoist infantry would be heavily supported by the Condor Legion's Junkers-52, Heinkel-111 and Dornier-17 bombers, as well as by fighters which machine-gunned the enemy's positions. Italian Savoia bombers, which the Germans considered superior to their own planes, added their quota of high explosive. Altogether, Franco's forces could count on some 163 aircraft, when the 73-strong Condor Legion was joined by a similar number of Italian warplanes including fresh Savoia SM-79 bombers to add to the S-81s already in service in Spain and CR-32 *Chirri* fighters sent from Italy on the orders of Mussolini as a first instalment of ninety aircraft with their crews, as well as by the aircraft now piloted by Spanish Nationalists. Airfields at Burgos, Logroño, Vitoria and Soria hummed with aircraft taking off and landing, being rearmed and refuelled. All the Francoist air forces available were brought together on aerodromes protected by the German 88mm artillery which operated in ground as well as in anti-aircraft mode.

On the Republican side, in October 1936, the Basque government bought four Gourdou-Leseurre GL-32 fighters, which the French air force had been using as trainers. Up to about seventeen GL-32s reached the northern Republican zone. On 30 April, six GL-32s, each armed with a 100-kilogram bomb, attacked the insurgent battleship, the 1912 vintage *España*, as it was trying to intercept a British merchant ship making for Castro Urdiales on the Cantabrian coast, and claimed to have scored hits. The *España* sank slowly while the aircraft attacked it. A British Admiralty investigation decided that the *España* had hit one of its own side's mines and had not been sunk by bombs, a view which, given its theory that aircraft could not score hits on moving warships, it hoped was correct.[22] In this case, it almost certainly was. The Republic's air chief, Hidalgo de Cisneros, writes that, given the limited weight of bombs that could be carried by the planes which were trying to scare off the *España*, they would hardly be able to sink the battleship.[23]

Spanish Republican emissaries had been desperately seeking warplanes in Europe. Eight Czech S-231 fighters, considered obsolete at the time, were sold to the Republic in the winter of 1936–7. After bureaucratic delays and the payment of sweeteners, given that the sale was illegal under the terms of the pan-European Non-Intervention agreement, the machines were unloaded at Santander on 17 March, assembled, fitted with armament – they came unarmed – and transferred to Bilbao. Between damage sustained in test flights, accidents and losses in combat, soon only one was left, to be captured when the northern port of Gijón was taken by the insurgents in October 1937.[24] Twenty-three Czech Aero-101 bombers were also bought, but the ship on which they were being brought to Spain, unescorted by the Republican warships which should have been detailed for the purpose, was intercepted by the Nationalist cruiser *Almirante Cervera*. The Czech machines were put into service with the Francoist air force.[25] Between April and June 1937, fourteen Dutch Koolhoven 51 and 51A light aircraft were flown to northern Spain via Catalonia and France. All were destroyed. A crew of two, intending to desert, had landed at Pau. In another one of the deals in which the Republican purchasing missions had to engage because no country could sell them arms openly, the Estonian government agreed to act as the ostensible buyer of aircraft from Czechoslovakia in return for the Spanish Republic buying its out-of-date material for a sum which would have paid for a squadron of British Spitfires if these could have been acquired. So, on 5 July 1937, eight Bristol Bulldog fighters – the plane in which the famous future RAF ace Douglas Bader crashed in 1931, losing both his legs – together with eight Potez 25s of 1927 vintage, were unloaded from under a cargo of potatoes at the port of El Musel (Gijón). These, and a heterogeneous assortment of other aircraft, known as the Krone Circus,[26] and some forty *Chatos* and *Moscas* would be left to defend the last of the northern zone of the Republic, against the Condor Legion's He-51s, Messerschmitt fighters and bombers.

On 5 May, four I-15 *Chatos* were destroyed on the ground at Sondica airfield near Bilbao. Only six Soviet 1-15 *Chatos* were left, reserved for the defence of Bilbao, so the Condor Legion's reconnaissance He-70s and He-45s could now roam at will over the Basque defences, taking detailed photos.

On the fine, bright morning of 31 March 1937, the 4th Navarrese Brigade, composed of enthusiastic volunteers of the Carlist or Traditionalist movement, known as *Requetés*, who marched with a fierce military chaplain bearing a crucifix at the head of their columns, moved out of their positions at 7:20 am. For the first time, a Nationalist infantry advance was going to go forward at several points on the 60-mile front in close coordination with the aircraft of the Condor Legion, dropping 50- and 25-kilogram bombs, accompanied by their fearsome 88mm guns (known to the Spanish Republicans as 'La Loca' or 'the madwoman' because of their speed and ferocity) and by some Italian aircraft. The Spanish infantry had white patches sewn into the backs of their tunics so that they would be identifiable from the air, while white arrows pointed towards enemy positions. Sperrle and von Richthofen were accompanied by General Velardi of the *Aviazione Legionaria* and by the Nationalist chief of staff for the attack, Colonel Vigón, as they trotted on horseback like nineteenth-century generals about to look over a prospective battlefield, although the techniques which the Germans were about to use were terrifyingly modern. As they waited, twenty Ju-52 bombers thundered overhead, and shortly afterwards they could be seen unloading their bomb loads on the positions of the *Euskogudarostea*, the Basque Army, or in Castilian, *Ejército de Euzkadi*, the Basque name for their provinces. At 1005 hours, the Spanish infantry moved forward as German He-51s and He-70 reconnaissance and ground-attack aircraft strafed the enemy a mere 50 metres ahead of the advancing men, and shells from the powerful 88mm artillery of the Condor Legion crashed into the Basque redoubts.

It was this concentrated bombing and shelling very closely in front of the attacking Nationalist infantry which forced the disciplined and well-fortified defenders out of their mountainous positions. The technique required very close ground–air liaison; specially detached German officers were required to stand side by side with the Spanish commanders of the attacking infantry. Despite this, progress was slow and relations between the Francoist leaders and Sperrle and even with the more diplomatic von Richthofen were not easy. The Germans seemed to demand immediate reactions from the Spanish soldiery such as might be expected of the highly trained professional and almost gladiatorial German storm trooper but not of the just as valiant though less highly trained Spanish infantryman. In the German view, Spanish officers did not seem to possess the same level of urgency or the same sense of timekeeping that was ingrained in the behaviour of their German counterparts.

After the jump-off at the end of March, General Mola's advance continued in the first days of April. At the same time, Nationalist aircraft attacked the defenders' airfields as far away as Santander and destroyed several machines on the ground. Resistance by defenders who were well dug-in and in mountainous territory was fierce, however, and Nationalist progress, despite heavy preparatory artillery barrages and bombing, was slow, hindered also by the bad weather typical of spring in the mountainous parts of northern Spain. The German impression, reflected in von Richthofen's diary, is striking and significant:

> The Reds retreat step by step with great morale and human losses but somewhat fewer material losses. They resist and defend themselves stubbornly, each step we advance has to be taken by force. Since some [Nationalist] infantry advance only when the enemy is not firing, and the Spanish artillery fires too late, slowly and badly to defeat the enemy, the burden of fighting falls on aircraft, which first demoralise the enemy by causing casualties, and keep them moving [back], remove their will to fight and impede supplies and movement. The effort falls on the forward batteries, whose shots force the enemy out of their trenches and disperse them as we advance. We can do all this because the enemy does not have an air force. The few fighters they have had have been shot down or destroyed on the ground.[27]

The Republican air force had reacted as best it could. The squadron of Soviet 1-15 *Chato* fighters which had been unloaded at Bilbao on 1 November 1936 had been joined in January 1937 by a further fifteen machines. Most were stationed at Lamiaco, a secondary airfield south of Bilbao; some were in Santander and others further west in Asturias.

German raids on the Basque capital, Bilbao, had begun on 4 January 1937 with a raid by five Ju-52s. The defending *Chatos* took off, led by 20-year-old Felipe del Río. Ju-52s could be attacked head-on from just above the line of sight of the German pilot. As the bomb aimer of one of them, Karl Gustav Schmidt, squeezed into his pod slung underneath the belly of the bomber, gritting his teeth against the freezing 150 mph wind, Del Río fired four streams of tracer into the Junkers. Having jettisoned the bombs, Schmidt and the observer Adolf Herman made a forced landing. The latter fired his revolver at a crowd of angry civilians who kicked and beat him to death, but a Soviet airman, perhaps the Russian commander, Boris Turzhanskii, who had been a volunteer in the Russian civil war and an outstanding pilot, landed his *Chato* nearby and rescued Schmidt. Turzhanskii was recalled to the Soviet Union in February 1937, to be succeeded by Felipe del Río, who became the squadron leader of the *Chatos* in the north in February 1937. Like Tomás Baquedano, José Riverola and Leopoldo Morquillas, who would be his successors in command of the squadron, del Río had been a pilot for a short time only. Some of these new and young pilots of the Republican air force had had civil licences since before the civil war but most had learnt to fly warplanes only recently in the first cohort which enrolled at the base of Los Alcázares in south-east Spain in November 1936. All would have been first-class combat pilots had they had sufficient training to convert to the *Chatos* with which they tackled the new Messerschmitt Bf-109 fighters and the meticulously prepared men of the Condor Legion as well as the pre-war Francoist pilots with their ample flying experience.

At the beginning of the eighty-two days from the start of the offensive until Bilbao fell on 20 June 1937, the Spanish infantry of the four Navarrese brigades and a mixed Spanish-Italian infantry brigade named *Frecce Nere* or 'Black Arrows' captured the important town of Ochandiano, gateway to the mountain passes and roads to Bilbao. There was little threat from the enemy's *Chatos*, the first of which to be downed by a Bf-109 fell on 5 April to the guns of Gunther Lützow, whose score in Spain was to be five and who would shoot down 103 enemy planes in the Second World War. On 18 April, however, one of the *Chatos*, flown by Felipe del Río, who would himself be killed a few days later, succeeded in shooting down a Dornier-17, killing its three crew. This was an important lesson, for the Germans had assumed that the speed of the Dornier 'Flying Pencil' rendered it safe from fighters. From then onwards, German bombers were escorted.

Despite what was, in terms of the Spanish Civil War, a steady advance, backed by immense air and artillery force, bombing and shelling the small Basque towns of Ochandiano, Elgueta and Elorrio, Sperrle tactlessly criticized Franco and his air chief, Kindelán, to their face. They were too slow. They were giving time for the enemy to build up his defences. Franco thanked them politely for their advice, but the wiser von Richthofen interpreted Franco's response as courteous prevarication.[28] On 18 April, frustrated because the Spanish infantry were not ready to advance, von Richthofen tried to reach the chief of staff, Colonel Vigón, who was tracked down to church where to the irritation of the German, it being Sunday, he was hearing Mass.[29]

The Condor Legion still had much to learn. The textbook attacks planned by Sperrle and von Richthofen did not have suitable failsafe plans on how to be flexible if the weather turned bad, as it did on many days in the rainy spring of the mountain Basque region. Despite the white patches on the backs of infantrymen's tunics, aircraft did sometimes drop bombs on friendly troops, which worsened the already tense relations between the Condor Legion and the Francoist command. And Spanish infantry were scarcely trained in moving forward behind a creeping artillery barrage or after low-level bombing of the enemy.

Bombing civilians

As the Condor Legion and its Italian allies flew over Franco's troops battling their way between 20 and 26 April towards the Basque capital of Bilbao, by which date the Germans had intended to have taken that city, they concentrated their efforts in the direction of a small town called Guernica, whose population was swollen by refugees. It was a road and railway junction, about 20 miles from Bilbao and about 10 miles from the swiftly moving front. The destruction of Guernica from the air would become symbolic of the Spanish Civil War and indeed of the terror bombing of civilians, a precursor of Rotterdam and Coventry.

Despite some heavily reported bombing of Madrid and of harbours such as Valencia and Cartagena, and occasional droppings of bombs by the Republicans on Zaragoza and Córdoba and other insurgent-held cities, air raids had not thus far taken the form of the mass slaughter of civilians and the devastation of large areas of cities which had been forecast by the British author H. G. Wells in his 1936 film *Things to Come* and by the followers of Giulio Douhet, the Italian advocate of mass terror bombing of cities as a way to force an enemy to surrender without being defeated in the field. The widely reported destruction of Guernica, including the filming of the ruins, which would be shown in cinema newsreels seen by millions of people all over the world, would bring the deep-rooted fears of bombing up to the surface. The Japanese had bombed Chinese cities. British, Italian and indeed Spanish aircraft had used bombs to suppress colonial insurrections, but Guernica was a European town bombed by West Europeans, which brought the horror much nearer home. Nevertheless, Guernica was not even the first heavy raid on civilians in Franco's northern campaign. As the assault on the Basque

Country began on 31 March 1937, German Ju-52s and two squadrons of Italian S-81 aircraft, with orders to destroy troops and military installations, bombed Durango, a town of ten thousand inhabitants and a road and railway junction between Bilbao and the front, with high explosive and fragmentation bombs, while fighters machine-gunned fleeing civilians.[30] Churches were hit during the celebration of mass, killing fourteen nuns and the officiating priest. Altogether, in this and two further air raids on 1 and 4 April, around 336 people died at once or of their injuries.[31] General Mola had proclaimed by radio and in leaflets dropped from aircraft over the Basque towns, 'If submission is not immediate I will raze [*arrasaré*] all Biscay to the ground, beginning with the industries of war. I have the means to do so.'[32] This was clearly an expression of ideological hatred rather than a tactic, but the threat had not led to the evacuation of non-combatants from Durango or to the building of air-raid shelters, which accounts for the immense loss of life. Yet the arms factory in the town remained intact and the roads through Durango were not blocked. Effective bombing evidently needed to be both more intensive and accurate.

The theory of total air war of this kind, unknown so far in Spain, implied destruction of the enemy's bases, lines of supply and troops in reserve. In von Richthofen's operations room in the Hotel Frontón in Vitoria, as recalled years later by Lieutenant Hans Asmus, a pilot attached to the Operations section of the Condor Legion, where the curtains were drawn for security and the leather-covered sofas and chairs were supplemented by long trestle tables strewn with maps, charts, weather and intelligence reports, a note over von Richthofen's signature was pinned to a board. Bombers were to attack military targets but to do so 'without regard for the civilian population'.[33] The reserve troops and the small-arms factory in Durango were legitimate military targets, but could they be accurately targeted without killing civilians? Did it matter? Should civilians be evacuated from places which constituted military targets? The Francoists, for their part, may be assumed to have had mixed feelings. Certainly they did not wish to destroy the resources of their own country, and even less did they want to drop bombs on their own supporters living in the Republican zone. On the other hand, General Mola, and one may assume he was not alone, wanted to destroy the factories, mines and railways of Basque industry because he saw it as part and parcel of the separatist movement which the military in particular hated. Furthermore, the Basques might well be fervent Catholics, unlike the burners of churches, murderers of priests and 'Godless Reds' of Madrid, Barcelona and Valencia. Nevertheless, by not fighting shoulder to shoulder with the equally fervent Catholic and neighbouring Navarrese Traditionalists and by allying themselves with Madrid for the sake of their autonomy, Francoist opinion thought that the Basques deserved to be crushed mercilessly.

Despite the presence of the Dean of Canterbury Cathedral, the Rev. Hewlett Johnson, known as the 'Red Dean' for his sympathy with the Soviet Union, who was visiting Durango when it was bombed again on 2 April, and had a lucky escape, the destruction of the town did not create the wave of international outrage that would become headlines in the world's press when Guernica was destroyed on 26 April 1937 by aircraft of the Condor Legion.[34]

Guernica

On the afternoon and early evening of Monday, 26 April 1937, the town of Guernica, with a population of seven thousand swollen to perhaps eleven thousand by civilian refugees and fleeing Basque militia, was destroyed with high explosive and incendiary bombs, the first time the latter, a mixture of powdered aluminium and iron oxide, had been used in Spain. In most people's minds, even today, after the bombings suffered by Amsterdam, Coventry, Hamburg, Dresden and Vietnam, Guernica remains the archetype of the destruction from the air of an undefended town for the purpose of terrifying its civilians. The news of Guernica was circulated by Reuter's press agency and the town was visited by leading journalists, among them George L. Steer, who wrote for *The Times* and the *New York Times*, the Belgian Mathieu Corman, who wrote for the Parisian *Ce Soir*, and Noel Monks, freelancing for the London *Daily Express*. Guernica was front-page news.[35] It broke on the Tuesday evening but became a major issue when Steer's dispatch appeared in *The Times* and the *New York Times* of Wednesday, 28 April.[36] Pablo Picasso's symbolic painting 'Guernica', which he painted in response to a commission from the Republican government for a work of his choice for exhibition that summer in Paris, later toured Great Britain, Scandinavia and America and was exhibited in the Museum of Modern Art in New York until several years after the death of Franco in November 1975, when it returned to Spain, where it constitutes a permanent reminder of Guernica's fate.

This was not the first air raid in Spain. Republican bombers had dropped bombs on cities in the insurgent zone, but against military targets and not as the Condor Legion would at Guernica, in massed formations and in repeated waves. The Republican action which caused most deaths occurred on 7 November 1938, when three Katiuskas dropped bombs in Cabra, province of Córdoba, killing 108 civilians. Republican information services believed that Italian forces were stationed in the town. Once over the target, the pilots assumed that the awnings in the market were army tents.[37] German aircraft had bombed Málaga and caused many civilian casualties, but they had done so in the process of destroying fuel stores. Madrid had also been bombed by Junkers bombers but mostly close to the front line which could be said to be a genuine military objective. While Guernica was a justified target, because it contained an arms factory – though this does not seem to have been targeted – and was a road and rail junction of some importance, its complete destruction by high explosive and incendiary bombs calls into question the motives of the Condor Legion and indirectly of Generals Franco and Mola and the chief of staff of the Navarrese brigades, Colonel Vigón. To what extent did the Nationalist command request or at least turn a blind eye to the total destruction of Guernica? That Francoist propaganda for many years denied that Guernica had been bombed and accused the retreating Basque forces of having blown up the town as they obeyed a sort of 'scorched-earth' policy suggests less that the insurgent command ordered the bombing in the form that it took, but that it went far further than they had requested and that they were highly embarrassed. What remains, however, is to investigate the tactics used by the Condor Legion, why the results appeared at the time to be unsatisfactory and, if possible, what responsibilities can be fairly stated to be those of the Nationalist military command.

Von Richthofen learnt from Condor Legion reconnaissance flights that Basque forces were retreating towards Guernica. This town was an important road and rail junction. The enemy could not be allowed to establish themselves in Guernica and hold up the advance to take Bilbao, the Basque capital. Condor Legion lieutenant Hans Asmus, attached to the staff, recalled looking at the map: 'All the roads joined together at Guernica, forming the artery that led across the bridge into the town. Then Captain Klaus Gautlitz, the Operations officer, standing with von Richthofen, drew a circle in red around the road junction and the bridge.' Von Richthofen earmarked the Rentería Bridge, over which the Basques would retreat, for destruction but, as Lieutenant Asmus recalled, ordered nothing to be done until, according to the agreement reached with the insurgent command, he had consulted Colonel Vigón, the Francoist chief of staff. Von Richthofen insisted that enemy morale was crumbling and that every effort should be made to complete their collapse. He later said that even a first-year army academy student could see what the target had to be.[38]

Though the bombing of Durango four weeks earlier had caused immense damage, it had not succeeded in blocking the roads for more than a few hours. Much more intensive bombing would be required. The bridge and crossroads at Guernica were to be hit, but in the Operations room in the Hotel Frontón in Vitoria, according to the memories of men questioned by Gordon Thomas and Max Morgan-Witts, nothing was said about the town of Guernica being only three hundred yards or so from the bridge and nobody asked about the consequences for this historic town and its people.[39]

So, did Colonel Vigón order or consent to the bombing, unaware or uncaring of von Richthofen's standing order that targets were to be bombed 'without regard to the civilian population?' According to von Richthofen's diary, he met Colonel Vigón early on the morning of 26 April, alone, for the German spoke good Spanish. But what was said at that meeting?

Von Richthofen ordered one of the newly arrived Heinkel-111s or perhaps a Dornier-17 to overfly Guernica. These planes, belonging to the experimental machines of the VB/88, were commanded by the highly skilled Rudolf von Moreau. Was von Richthofen intending to try to concentrate the bombs on the bridge? Could von Moreau reproduce his feat of the previous year of dropping food parcels into the sixty square yards of the inner courtyard of the Toledo Alcazar, the infantry academy in which 1,200 civil guards with women and children had been besieged by Spanish republican militia?

Von Moreau dropped twelve 50-kilogram bombs which hit a hotel, the railway station and the plaza in front of it. The first bombs failed to hit the bridge, but they did cause much damage and many deaths and injuries. Was this part of the bombing tactic against the town itself, or does it suggest the opposite, that is, that the bombing simply missed its target? Von Moreau then flew back to rendezvous with the rest of his squadron. Then three Italian Savoia SM-79s, which had left their base at Soria at 1530 hours to cooperate with the Germans, bombed the bridge from a height of 10,000 feet. The Italian bombers dropped thirty-six 50-kilogram bombs which fell around the bridge and the railway station and destroyed several buildings but failed to hit the bridge itself. Next, two Heinkel-111 bombers also failed to hit the target. Their mixture of high-explosive, shrapnel and incendiary bombs fell along a kilometre-long line from the bridge doing untold damage to the market and to shops and factories, and killing and maiming passers-by. Later, He-51 and CR-32 fighters machine-gunned fleeing people in order perhaps to force them to stay in the town centre. But not till about 1830 hours did the large bombing fleet appear. These were three squadrons of Ju-52s, a total of nineteen bombers of the 1st (von Knauer), 2nd (von Beust) and 3rd (Krafft von Delmensingen) squadrons, accompanied, despite the complete absence of enemy fighters, by ten Italian Fiat and German Bf-109 fighters. The bombers dropped at least 31 metric tons of high explosive, in a combination of 250- and 50-kilogram bombs, as well as a total of 2,500 incendiary bombs. The bombs created a huge cloud of smoke and dust, as was to be expected. The Bf-109s accompanying the bombers machine-gunned people fleeing, which must have been their purpose, given that there were no enemy fighters in the area. Guernica had no anti-aircraft defence, so the bombers targeted the area from a low height, destroying 74 per cent of the buildings. Arguments about the loss of life have continued ever since the outrage, with a maximum figure claimed of 1,645 dead and several hundred injured. While several air-raid shelters had been prepared by the local authorities, direct hits were responsible for many of the deaths. Because fires were still burning, it is likely that many bodies were not recovered before the Francoists entered the town. While the first figures, issued by the Basque authorities, are now thought to be an overestimate, the figure of between two and three hundred dead is generally accepted.[40] It may well be, however, that numerous later deaths in hospitals in the rearguard arising from the bombing, and corpses found long afterwards when the rubble was cleared, have not been included in the total mortality figures. They may not have been registered or they were buried elsewhere with no evidence of the cause or place of death.[41]

One major question is whether the Germans themselves wished to launch a violent attack on Basque culture, because Guernica had particular historic significance in Basque history. For centuries, Basque leaders had met under its famous tree and Spanish monarchs swore in the church of Santa María to uphold the independence of the historic rights of the Basque Country. However, von Richthofen did not need to know anything about Basque culture if he wanted to cow the enemy. The destruction of Guernica 'astonished' the chief of staff of the Condor Legion as his diary entries show. One suggestion is that '[the] divergence between von Richthofen's intentions and the way they were executed strongly indicates that the orders were garbled in transmission or upon arrival'.[42] The descriptions of the precise orders given do not, however, support this theory. Loading 2,500 incendiary bombs to be dropped on a town mostly built of wood cannot be ascribed to misunderstanding. Nor can the machine-gunning of civilians, unless it can be alleged that soldiers were among them.

From Germany repeated enquiries were made as to who had bombed Guernica, a fair question considering that General Sperrle had repeated the Francoist propagandist statement that the Basques had blown the town up in retreat as part of a scorched earth policy. Von Richthofen concluded that smoke had concealed the bridge, which had not been destroyed.

> The attack was with 250 kg bombs with fire bombs making about 1/3 of the load. When the first Junkers arrived there was already a lot of smoke (from the experimental squadron, which attacked with 3 aircraft). Nobody could recognise the streets, bridge and suburbs and *therefore* [they] *just dropped bombs anywhere*. The 250 kg bombs destroyed a number of houses and the water supply lines. Now the firebombs did their work. The kind of buildings: tile roofs, wooden beams, and wooden galleries, were the reason for the complete destruction. [. . .] The city was completely closed off for at least 24 hours; that would have guaranteed immediate conquest if troops had attacked right away. But at least it has been a complete technical success, with our 250s (bombs) and EC.BS (incendiaries).[43]

However, one might allege that German rehearsals and study would have made it obvious that fires started by high explosive and incendiary bombs create smoke and dust, which conceal specific targets. Nevertheless, von Beust, one of the leaders of the raid, reported as follows:

> [Because] the explosions of the first flight of bombers produced an enormous quantity of smoke and dust, the area of the bridge and the town was

> completely hidden from view, so that the next two flights could do no more than drop their bombs by rough approximation. Because of this, and because of a strong change in the wind, the bombs fell on the town.[44]

The Ju-52s lacked advanced bomb-sighting equipment. A pot looking somewhat like a dustbin was rigged in flight under the belly of the aircraft. The navigator/bombardier crawled into it and aimed through a simple bombsight, communicating with the pilot through push buttons which illuminated coloured lights in the cockpit, red for left, green for right and white for straight on. Von Beust's later comment was 'it's a wonder we hit anything at all'.[45]

To return to the responsibility of the Spanish Nationalist commanders, in the case of a mission being urgently required, as it indeed was on 26 April 1937, they probably did not go into detail, merely requesting bombing of the bridge and the roads. The unanswered question is whether the Francoist command led the Condor Legion to understand that it was permissible for the carpet bombing which the Germans were indeed studying to be used on an inhabited town. For the Condor Legion, the small towns in the Basque Country, built largely of wood, offered, in the words of a report of 28 May 1937, 'a type of construction similar to those in the small towns of Germany's neighbours'.[46] Was this a mere comment or could it be interpreted as a statement of future war aims?

While, during the latter part of the Franco regime itself, it became impossible to maintain the lie that the Basques had destroyed Guernica themselves, theories were publicized to try to remove the blame from the Francoist command and lay it entirely on the Condor Legion. From von Blomberg, the War Minister in Berlin, came demands for an official explanation, because the German authorities were faced with a diplomatic crisis caused by George L. Steer's dispatch in *The Times* of Wednesday, 28 April, and the confirmation of it by the British consul in Bilbao, and needed urgently to advise their diplomatic representatives on how to reply to enquiries, including the embarrassing situation in which the German ambassador in London found himself in the weekly meeting of the Non-Intervention Committee[47]. The reply from Vitoria was that it was not Germans who had bombed Guernica.[48] Given that this was well-known to be untrue, it was probably Franco's diplomatic advisers who concocted the following advice:

> Kindly request *Sander* (Sperrle) to tell Berlin that Guernica [...] is an important road and railway junction, with a bomb, pistol and ammunition factory. On 26 April enemy forces were withdrawing through Guernica and reserves were stationed there. Our front-line units requested air force to bomb the road junction. This was done by German and Italian aircraft. Owing to poor visibility, smoke and clouds of dust, the town itself was hit by bombs [...] the Reds took advantage of the bombing to set fire to the town.[49]

The message continued by referring to 'constant' bombing of cities in the Nationalist rearguard and to the murder of large numbers of civilians by 'the Reds' in the Republican zone. So, while admitting that the bombing, carried out by the Germans, was requested by the Nationalist command, the Franco regime claimed and went on claiming, almost to its end, that the Basques had burnt their town and that in any case the murders of civilians in the Republican zone somehow excused what had happened to Guernica.

The Condor Legion pilots themselves were ordered to say nothing about and even to deny the bombing.[50] Nevertheless, the fighter pilot Harro Harder wrote in his diary a moving text, which presumably would have caused him serious trouble if it had been discovered.

> Today we were in Guernica. It is completely destroyed and not, as the newspapers here say, by hordes of red incendiaries, but by German and Italian bombing. We all think it is a scandalous and foul deed (*'eine unerhörte Schweinerei'*) to destroy a militarily irrelevant town like Guernica.[51]

The journalist George Steer recalled driving later into Guernica later on the 26 April with other journalists who had been with him in Bilbao. Fires flamed through every window; the streets were scattered with beams, rubble and telephone cables. There were dead and limbless corpses everywhere, inside and on the edges of the bomb craters. Injured, homeless and exhausted women lay on rough tables and mattresses soaked by water from firemen's hoses, trying to find suitable words to tell the newsmen what they had experienced, bombs falling, bursting, houses falling, tubes of incendiary bombs bursting and flames spurting. Wreckage was crashing down all the time.[52]

The journalists drove back to Bilbao to send their stories. On 28 April, the British consul in Bilbao, Mr Stevenson, wrote to Sir Henry Chilton, ambassador to the Spanish Republic and domiciled just over the French frontier in Hendaye, the following brief message. For all its appropriately cool tone, his dispatch appalled the officials in London who read it:

> [Nine] houses in ten are beyond reconstruction. Many were still burning and fresh fires were breaking out here and there, the result of incendiary bombs which [...] had not exploded on impact and were doing so [...] under falling beams and masonry. The casualties cannot be ascertained.[53]

Now, more than forty years after the end of the Franco regime, the justifications which were given for the bombing, which were aimed at suppressing discussion about it, can be seen for mere propaganda. The Republicans had done nothing like Guernica and, before it, Durango. Nor is there any evidence of a 'scorched-earth' policy on the part of the retreating Basque forces. Last, although murders in the Republican zone were plentiful, they were almost certainly fewer than in areas controlled by the insurgents during the war, leave alone after it.[54]

The first aircraft to drop bombs on Guernica was the a Heinkel-111 or possibly a Dornier-17. When it, the Savoia S-79s and the He-111s all failed to hit the target, some time passed before the Ju-52s were dispatched. It would seem that, if indeed the first bombings were aimed solely at the bridge and crossroads, then, everything else having failed, a mass attack could not do any worse and, even if bombs from nineteen aircraft would obviously destroy large parts of the town, they would, hopefully, destroy the bridge also.

Why, given the absence of anti-aircraft or fighter defence at Guernica, did the detailed German orders not specify bombing the bridge from the lowest height possible? Why indeed did von Richthofen not experiment with his Henschel-123 dive-bombers? Perhaps they were not considered steady enough in a dive.[55] Yet none of the other bombers managed to hit the bridge. Did the Ju-52s have any means of asking for orders, given that the targets were not visible, or did they have orders to dump their bombs over the general area? In later years, pilots recalled that they had to drop their bombs because of the danger of landing with them still in the bomb bays.[56] However, this could have been done in open country or, had fuel reserves allowed, over the sea.

On Thursday, 29 April 1937, three days after the bombing, Francoist troops entered Guernica. The terrified survivors dared not say to the accompanying journalists that their town had been destroyed and their families slaughtered by bombers.

The aim of the Condor Legion was to produce a rapid victory for Franco's forces (especially in contrast with the Italian disgrace at Guadalajara just a few weeks before Guernica). And, as von Richthofen wrote, 'In practice we directed the whole business without having any real responsibility.'[57] Certainly, it could be said that the destruction of Guernica was 'the logical consequence of this policy of managing terror that Franco and Mola had embraced' and that Guernica reflected German theories of total war.[58] But whether that was the intention of the particular event is unclear. What Colonel Vigón understood was to happen when he discussed bombing Guernica with von Richthofen early on 26 April is the ultimate question, and, if he did know the answer, did he seek higher authority? The answer is probably in the negative, unless documents have been destroyed and German reports doctored. It is true that the Condor Legion could not act without the consent of the Spanish authorities, but even if Mola and Franco were informed of the intention to drop bombs on the bridge at Guernica, there is no evidence that the Germans intended to destroy the town and told the Spanish leaders that in so many words. On the other hand, the bombings of several towns on the previous day, with large numbers of houses destroyed and civilians killed justify perhaps the speculation that Franco, Mola and Vigón were not averse to that type of air warfare.

Throughout April 1937, Italian and German aircraft had regularly bombed Bilbao, the capital and major port of the Basque Country. The combined effect of the bombing of Bilbao, added to the horror of the destruction of Durango and Guernica, against the background of the Royal Navy being ordered to defend British-registered merchant ships from Franco's naval blockade, and in the context of a current of British public sympathy for the Basques as a people with similar values, led to pressure to admit four thousand Basque children to Britain as refugees.[59]

Chapter 7

The fall of Bilbao; the end of the war in the north of Spain: The two sides reorganize their air forces

As, in the spring of 1937, Francoist forces neared Bilbao, German aircraft constantly reconnoitred the defences of the Iron Belt or *Cinturón de Hierro*, the elaborate dugouts, bunkers, trenches, barbed wire and strong points built to defend the city.[1] In May and June 1937, the Condor Legion's bombers were active, while its He-51 fighters had copied the *cadena* ('chain') tactics invented by the Spanish Nationalist pilots in which formations of aircraft attacked in succession until their ammunition was exhausted. So many aircraft were available that the Francoists and the Condor Legion were able to maintain a sort of shuttle, returning to base to load bombs and ammunition.[2]

The government of the Spanish Republic tried to send aircraft from its main territory in central and eastern Spain to reinforce the northern front. Hidalgo de Cisneros, commanding officer of the air force of the Spanish Republic, writes that the problem was 'a real nightmare' (*una verdadera pesadilla*). The slightest change in wind or en-route diversion exposed the planes to falling into enemy hands.[3] On 8 May 1937 nine *Chatos* and six *Natachas*, accompanied by a DC-2 with ground crews, left Reus in north-eastern Spain to fly to Bilbao. They could not fly the direct route, which would take them through skies patrolled by enemy aircraft, so they landed at Toulouse. Here the Spaniards came up against internal French tensions. It had been agreed that the machines could refuel in France, but, shortly after they had landed, the Dutch colonel Lunn, local officer overseeing the functioning of the Non-Intervention Agreement, accompanied by a French officer, set guards around the *Chatos* and the *Natachas*, and interned the Spaniards.[4] That night the Spanish crews were given hospitality by French families. Finally, they were allowed to leave but told that they must go to the Republican capital, at the time Valencia. The firing pins of their machine guns were removed and a French fighter squadron escorted them to the Catalan frontier. On 17 May, a similar fate befell another squadron of *Chatos*, together with three *Natachas* and an escort of two DC-2s, which landed at Pau. Only four planes out of another expedition of nine I-16 *Moscas*, leaving the airstrip of El Soto at Algete near Madrid at the end of May, reached La Albericia airfield in Santander in the northern Republican zone. The others lost their way and ran out of fuel.[5] Anguished telegrams sent to Prieto, the Republican Minister of Marine and Air, by José María de Aguirre, the president of the Basque region (and ex-footballer of the Athletic Bilbao club, founded by British steel and shipyard workers and Basque students returning from schools in Britain), between the beginning of the campaign on 31 March and 13 June, reveal constant demands for aircraft. Aguirre insisted that *Katiuska* bombers could land at the two airfields of Lamiaco and Sondica, near Bilbao, and at La Albericia (Santander). On 16 May, he cabled that the distance between Guadalajara and Reinosa (Santander) was 270 kilometres, implying that aircraft could be sent from the central front to the north. A few days earlier he had said that the commander of the Republican air force, without mentioning names, was suspected (without saying of what). The last telegram before the fall of Aguirre's capital, Bilbao, read,

> There are no troops, only solution is aircraft in mass to appear today. If not, Bilbao will fall in hours.[6]

Why were more Russian aircraft not sent to the northern front? This is one of the unsolved questions of the Spanish Civil War. Were the Soviet commanders unwilling to risk losing valuable machines in trying to cross enemy-held territory? Certainly, the risk was real. Only a few of the *Chatos* and *Moscas* which attempted to reach the northern zone succeeded. Julián Zugazagoitia, parliamentary deputy, editor of *El Socialista* and intimate of Prieto, Minister of Marine and Air, and who would in 1940 be sent back to Spain from France by the Nazis, and executed by the victorious Franco, writes that Prieto did his best to send aircraft to the north but does not explain why he did not.[7] One reason may be that the inexperienced Republican pilots were overwhelmed by the Italian professional flyers. Only one Fiat *Chirri* was lost in the entire battle for Bilbao.[8] This was on 5 June. After escorting Savoia S-81 bombers in a raid to destroy artillery in the 'Iron Belt' around Bilbao, seven *Chirris* had stayed to cruise around the area in search of prey. They discovered an airstrip from which *Chatos* were about to take off. Diving down, the Italians destroyed four Russian fighters on the ground and set fire to the fuel stores. However, one *Chato*, flown by Rafael Magriñá Vidal, another of the ex-sergeants of García Lacalle's squadron who had started his flying training only the previous November and who would be shot down and killed in August 1937, shot down a *Chirri* piloted by the highly experienced Guido Presel, who had had thirteen confirmed victories in Spain. Possibly his guns jammed at the vital moment or he had exhausted his ammunition.[9]

The Condor Legion poured all its resources into the saturation bombing of the Iron Belt. Every night the ground crews cleaned and reloaded guns, refuelled planes and serviced engines. During the day aircraft would land, be reloaded and refuelled and back in the air in a few minutes. In his reports, von Richthofen emphasized the contribution of the German aircraft and particularly of the 88mm flak artillery used to eject the enemy from their positions.[10] By June 1937, German air–ground coordination had greatly improved. There were far fewer incidents where friendly troops were strafed. The final assault on Bilbao began on 11 June. Waves of aircraft attacked defensive positions. Constant air raids paralyzed industry, while families spent the days outside the city or in railway tunnels. The Basque capital, blockaded, defenceless, starving, bombed and shelled, could no longer be defended. On 19 June, the Basque forces retired westwards into Santander province and Franco's troops marched into one of Spain's larger cities.

The next offensive, westwards along the Cantabrian coast towards Santander, began on 14 August 1937. Finally, to wipe out the shame of their defeat in the Guadalajara offensive of March, the *Corpo di Truppe Volontarie* – the Italian corps sent to aid Franco's Nationalists – had been reorganized. Five thousand men had been repatriated, leaving the most suitable volunteers to fight side by side with Franco's Navarrese brigades and one brigade, called *Frecce Nere* ('Black Arrows') of mixed Italian and Spanish troops. In the battle for Santander, between 14 and 26 August 1937, the *Aviazione Legionaria* put up twenty-seven S-79 and S-81 bombers, forty-one CR-32 *Chirri* fighters and twenty-four Imam Romeo-37 two-seater fast reconnaissance biplanes, equipped with cameras and wireless transmission facilities. The Italian air force called them *lince* or 'lynxes' but in Spain they were always called *Romeos*.[11] He-111 bombers escorted by Messerschmitt Bf-109 fighters flew into enemy territory, bombing one vital target after another. They were intercepted by Soviet I-15 *Chatos* and odd aircraft from the heterogeneous 'Krone Circus', which were regularly lost to the superior Bf-109s and the highly trained pilots of the Condor Legion. No Condor Legion aircraft were lost. In one case, the anti-aircraft guns of the Francoist cruiser *Almirante Cervera* managed to down a *Chato*.

Some ten *Chatos*, protected by I-16 *Moscas*, had managed to make their way to the northern zone that spring and summer. Four more, led by Alexander Senatorov, flying an SB *Katiuska* to guide the pilots, newly returned from training in the Soviet Union, were sent on 18 August, protected by two squadrons of I-16 *Moscas*, led by Boris Smirnov. This Russian, together with three other *Mosca* pilots, was shot down and hospitalized three days after his arrival. He was replaced by the ex-sergeant pilot, now captain, Manuel Aguirre. A further seven

Moscas out of a squadron of twelve which had left the central zone, flying directly to Santander, arrived at their destination on 2 July. They were commanded by Lieutenant Valentin Ukhov.

The Republican pilots who on 21 and 24 May 1937 attempted to fly from airfields around Madrid over the mountain ranges as far as the northern front often found that they had to divert from the direct course. On the second of these dates, when only four out of nine fighters reached Bilbao, Italian Fiat CR-32 *Chirris* reported having forced José Bastida, another ex-mechanic relatively newly qualified as a fighter pilot, to land just off the beach in the sea at San Sebastián, in enemy territory. However, Bastida himself, years later in exile in Mexico, told his ex-commander Andrés García Lacalle that he had merely run out of fuel. He sat on his wing until fishermen picked him up. He was fortunate, because, having been an air force mechanic, a Francoist court martial could have charged him with mutiny as well as the accusation of military rebellion which was the general crime which allowed military courts to condemn men who fought for the Republic. In the event, he was exchanged for Italian pilots in Republican hands. In later years, Bastida would run a highly successful flying school in Mexico.[12]

On the northern front, the senior Soviet adviser to the Spanish Republican colonel Antonio Martín-Lunas, who had led the international squadron in which the British pilots Cartwright and Oloff de Wett had flown, was Colonel Fedor Arzhenuknin.[13] Against their mere forty-five machines, the Condor Legion could put into the air a squadron of Messerschmitt Bf-109s, two of He-51s, two squadrons of Junkers-52s and one squadron of He-111 bombers, together with Henschel 123 dive-bombers and He-46 and H-70 reconnaissance aircraft, a total of about seventy operative machines, while the Italian *Aviazione Legionaria* had five squadrons of Fiat CR-32 *Chirris*.

In April 1937, the *Aviazione Legionaria*'s CR-32 units were reorganized and increased in size. I and II groups were disbanded, redesignated and replaced by two groups that each controlled three squadrons as before. A third new group, the VI, was also formed.

The XVI fighter group (a wing in RAF terms) under the command of Major Giuseppe Casero (*Casetti*) included the 24th, 25th and 26th squadrons. It adopted the name *Gruppo Cucaracha*, after the popular Mexican song of the period and used as its insignia a winged Moroccan cockroach which was applied to the sides of the group's aircraft.

The XXIII fighter group was formed on 22 April under the command of Major Andrea Zotti (*Biondi*) and included the 18th, 19th and 20th squadrons. They called themselves the *Asso di Bastoni* (ace of clubs) and its CR-32s were adorned with a marking adapted from Neapolitan playing cards showing a weapon used by the Fascist action squads.

To these forces were added the German and Italian aircraft now being flown by Spanish Nationalist pilots. The Francoists had overwhelming superiority in aircraft, logistic back up and pilots, for the Russians were being withdrawn and it was newly trained Spanish Republican pilots, just back from six months' training in the Soviet Union, who had now to engage in combat with highly trained German and Italian military pilots.

Santander fell on 26 August. Of those last weeks, as the Republican pilot Francisco Tarazona writes:

> Our missions become ever more frequent. We have 16, 14, sometimes only ten machines ready to fly. We have to machine-gun the enemy and, protect our own lines. Physical exhaustion overcomes us.[14]

Next, Francoist forces advanced on Gijón, a port and city in Asturias, in an offensive starting on 1 September and ending on 21 October 1937. In this campaign, in which the miners of the coal basin of Asturias fought bitterly, the Condor Legion struck communications, concentrations of men and vehicles, mountain passes, ports and aerodromes. He-51s, flying several sorties daily, were used with devastating effect for close support of infantry. Each movement was planned with Franco's staff officers. The Condor Legion assigned liaison teams to front-line units. Although this tactic steadily improved, there were still some cases where Francoist infantry were strafed by friendly Condor Legion fire, and the radios in service were still found to be inefficient at air–ground communication.[15]

By early September, the whole of the Condor Legion's J/88 fighter force had assembled at one aerodrome, from which the Bf-109s escorted bombers striking at Gijón, while squadron 3.J/88 flew more freelance patrols, operating directly in front of the advancing ground troops, as well as engaging in the art of *Kochenjagd* ('hunting to eat') – a nickname given to vehicle-hunting missions – often at dusk. Adolf Galland remembered that during such intensive operations in the He-51:

> reloading the machine guns you usually cut your knuckles open on one of the many obstacles in the unbelievably confined space of the overheated cockpit. On hot days we flew in bathing trunks, and on returning from a sortie looked more like coalminers, dripping with sweat, smeared with oil and blackened by gunpowder smoke.[16]

On 21 September, the 3rd German squadron moved to the airfield at Llanes, on the Asturian coast, which had only recently been captured. From here, it harried the Republican troops as they retreated. Adolf Galland recalled:

> Llanes was funniest aerodrome I have ever taken off from. Situated on a plateau whose northern side fell sheer into the sea, with the three remaining

> sides almost as steep, it was like taking off from the roof of a skyscraper situated on the seashore.

Captain Gotthardt Handrick recalled the combats over Gijón:

> Our raids on Gijón did not always go quite so smoothly and without problems. The Reds also put up enormous resistance against us in the air, and their *Ratas* and *Curtisses* were by no means easy prey. During the course of one air battle, I was engaged intensively with a Curtiss, and since it was one of my first air battles, I was naturally convinced that I would be able to take care of the enemy crate in an instant. Far from it! Apart from the fact that my machine guns were not firing well for some reason or other, I must admit that I had not handled my position too well. I attacked the Curtiss from behind – it did an about-turn. Instead of pulling my nose up and then repeating the attack, I immediately attacked again, and at each attack managed to fire only ten rounds, whilst the Curtiss, which, of course, was slower than my aircraft, was able to compensate for that by being very manoeuvrable. It also had an excellent rate of climb and the pilot shot well.
>
> Gradually, we distanced ourselves ever more from the coast, and in the end we were a good five kilometres out over the sea, directly facing Gijón, the front being about 50 to 60 km. away. Hence it was high time to call it a day! I drew closer to the Curtiss, ever closer, so close that in the end there was a weighty thud – I had collided with it.
>
> On my aircraft, the wing had been hit directly at the fuselage root. In addition to that, the controls jammed. I made three or four involuntary rolls, one after the other, until I was finally able to flatten out the crate and with much effort fly in a straight line. Should I bail out, over the sea? No thank you! The water was too cold in October! I flew inwards towards the land. If someone were to catch hold of me now – I dared not bring this thought to a conclusion. At last I arrived over the front, and then over Llanes. Here, over the airfield, I would be able to fling down the aircraft, but the airfield commander will berate me. 'Where in the world do you think we can obtain spare parts when you fling the crate down? How are we to transport it away from here? All the bridges in the hinterland have been blown up, and it can take weeks before you'll have a new aircraft'.
>
> Under similar comforting thoughts, and with a choking (sic) aircraft, I reach Santander. Thank goodness! I'm at a height of only 150 metres. Very cautiously I let the undercarriage down. In doing so, out of my aircraft a piece of the wing of the enemy Curtiss flies away – a souvenir of one of my first air battles which I keep even today. After landing, I took a look at my bird. It had been badly damaged. The right wing was a mess, the fuselage was dented in, at the back were a few hits and the empennage needed to be repaired.

> However, I was very happy that I had been able to hold out, for after two days the damage had been more or less taken care of – and the bird continues to fly.[17]

The battle for Gijón saw dogfights between Bf-109s and Soviet I-16 *Moscas*. It took several weeks to reduce the fierce defence of Asturias. After several German aircraft had been shot down, the Condor Legion had to rethink its tactics. One of the developments was an early version of what would later be called 'Napalm'. It was a petrol can filled with fuel and oil and with a fragmentation and an incendiary bomb attached.[18] On 27 September 1937 Harro Harder, now a captain, had seven victories in Asturias. His total in Spain would be eleven, followed by a further eleven in the Second World War. However, the superior training of the Germans and perhaps the quality of their aircraft told, and no Bf-109s were downed. In contrast, twenty-one I-16 *Moscas* were lost on the northern front between 12 August and 21 October 1937. Most were piloted by recently trained Spanish pilots, though the Russian commander Boris Smirnov crash-landed on 19 August and parachuted from his machine on 24 August. The Basque squadron, formed by the I-15 *Chatos* which arrived at Bilbao in November 1936 under the command of 'Bores' (Boris Turchanskii) and then under Felipe el Río until he was killed on 22 April, was wiped out in the campaigns in the north. The Spanish Nationalists and the Condor Legion claimed to have shot down or otherwise destroyed fifty-one aircraft. The city and port of Gijón finally fell on 21 October. The entire north of Spain was now in Franco's hands and he could redirect his Italian and German air forces to other fronts. In all, forty-eight aircraft fell into the hands of the Nationalists, among them the twenty-two Czech Aero-101s which had been captured by a Nationalist ship while being shipped to Spain.

As might be expected, German commanders sent back detailed studies and reports of the campaign to *Sonderstab W*. Hitler now recalled Sperrle. His rudeness to the Spanish commanders had been embarrassing. On 30 October 1937, Sperrle turned his command over to General Helmuth Volkmann, a fighter pilot in the First World War, whose nom-de-guerre on taking over the Condor Legion would be 'Veith'.[19]

The two Spanish air forces are reorganized

From 8 August 1936, insurgent aircraft were repainted with white tailplanes and a St. Andrew's Cross. General Alfredo Kindelán, one of the founders of the Spanish air arm, a previous commander of the air force and a prominent monarchist who had retired when the Republic was declared in 1931, was in the southern seaport of Cádiz, acting as General Mola's link with the officers of the Spanish navy who had agreed not to blockade the ports from which the Spanish forces in Morocco were to sail to the Peninsula. Franco appointed Kindelán to command the various parts of the air force which were in rebellion against the Republic.[20] Kindelán in turn was one of the strongest advocates of handing Franco supreme political as well as military power.[21] Among decrees issued by the embryonic insurgent government, known as the National Defence Junta, was one appointing as chief of air staff, Lieutenant Colonel Sáez de Buruaga, a senior staff officer. In 1936 he held no appointment but joined the uprising in the León base, after which he was placed in command in the north. Previously, on 18 August 1936, exactly one month after the beginning of the military insurrection, a number of decrees appeared in the *Boletín Oficial del Estado*, the official gazette of the insurgents, laying down some basic organization which would fit the particular circumstances of the war and the material and personnel available. There were to be three fronts: North, Central and South, and then a fourth for the East (Levante) with its headquarters at Zaragoza. Altogether there were about one hundred airfields, a figure which would double as the Nationalists advanced into Eastern Spain and Catalonia in 1938. The Italian aircraft and pilots were fictionally incorporated into the Legion. The Fiat CR-32 *Chirris* became the 1st Fighter squadron nicknamed, as has been seen, *La Cucaracha*, while the Germans remained under 'Operation Magic Fire' or *Unternehmen Feuerzauber*. A few of the crack Spanish pilots such as Joaquín García Morato, Angel Salas, Julio Salvador and Miguel García Pardo flew with *La Cucaracha* sometimes, while the slower Heinkel fighters and reconnaissance machines were handed over to the Spaniards. The insurgent air force, called *Aviación Nacional* by the Francoists, began its organization at the end of September 1936 once Franco had been appointed to supreme command as well as to political leadership of the insurgent part of Spain. In November 1936, the newly arrived Condor Legion, the Italian *Aviazione Legionaria* and the Spanish *Arma de Aviación* became three distinct entities. More and more Spaniards took over as trained crews as aircraft became available. Fighter groups of He-51s, CR-32s and bomber groups of Ju-52s, Savoia-Marchetti S-79s and S-81s, plus advanced German He-111 and Dornier-17 bombers, were organized in groups according to number and type, producing groups numbered, for example, 5 G (Group) 28, corresponding to S-79 Italian bombers.

In November 1937 constant changes in commanders, organization and reorganization of groups (*Escuadras*), wings (*Grupos*) and squadrons led to the construction of the 1 Air brigade or *Brigada Aérea Hispana*, under Sáez de Buruaga and Major Arranz (who had flown to Berlin with the two German civilians to secure aid in July 1936). The purpose was to increase combat efficiency. The 1 Air brigade included six bomber wings, two fighter wings and one of reconnaissance aircraft. The II Air brigade had four groups, the bombers with two wings each and the fighters with four groups.[22] Although there were also independent groups and some independent squadrons, the idea was to regroup all the existing bomber wings and to emulate the tighter structure of the *Aviazione Legionaria* and the Condor Legion. Behind the thought of the overall Nationalist air force commander, General Alfredo Kindelán, was the doctrine of absolute dominion of the air in a war, an idea formulated by theorists such as the Italian Giulio Douhet and the British Hugh Trenchard after the First World War.

Commanders of Francoist air force units, unlike their Republican opposite numbers, were all men who had been officers before the war. The major regions were commanded by Captain Julio García de Cáceres, Majors Roberto White, who had commanded the aerodrome at León, Julián Rubio, commanding the aerodrome at Logroño, and José Rodríguez and Díaz de Lecea. On the insurgent side there was an obvious disinclination to promote officers, although the practice of temporary upgrading (indicated by a black patch with insignia of the higher rank) was common. This contrasts greatly with the generalized 'loyalty' promotions on the Republican side.

Here, the arrival of Francisco Largo Caballero's government in early September 1936 led to the creation of a ministry for the navy and the air force (*Ministerio de Marina y Aire*) separate from the War Ministry under Caballero himself. The new ministry was headed by Caballero's fellow socialist Indalecio Prieto, who placed Lieutenant Colonel Ignacio Hidalgo de Cisneros, who had helped Prieto to escape from Spain after the failure in 1930 of a Republican plot, at the head of the air force of the Republic. The Undersecretariat for Air was placed in the charge of Lieutenant Colonel Angel Pastor, one of the few senior air force officers to remain loyal and who had been temporarily in charge of the air force in the first few chaotic weeks when the attitudes and political loyalties of air force personnel were uncertain. Pastor, who himself seems to have been in danger because of unjustified political suspicions, was soon sent abroad on a mission to acquire aircraft, as was another officer with the same patronymic and probably insurgent sympathies, Major Carlos Pastor Krauel.[23] Another Spanish air force officer, Captain Agustín Sanz Sainz, whose loyalty was not in doubt because he had fled Spain for Venezuela and the United States after refusing the orders of Franco, then chief of staff, to bomb rebellious miners during the uprising in the mining districts of Asturias in October 1934, offered to help the Republic buy aircraft in the United States. Sanz Sainz would be killed on 16 July 1937 when six Savoia-Marchetti S-79s bombed the air base at Alcalá de Henares, which he was commanding at the time.[24] He was joined by yet another senior Spanish air force officer, Colonel Francisco León Trejo, commander in 1936 of the Cuatro Vientos air base near Madrid, while Lieutenant Colonel Juan Ortiz, who had saved the Los Alcázares air base for the Republic in July 1936, and who had had a serious disagreement with the Soviet advisers, was sent to Paris to buy aircraft.[25] Colonel Angel Pastor Velasco was soon replaced as Undersecretary for Air by Lieutenant Colonel Antonio Camacho, who had commanded Escuadra I, based at Getafe. Major Angel Riaño Herrero was named chief of air staff under the command of Hidalgo de Cisneros.[26] Thus many Spanish Republican senior officers, of whom there was no shortage, were not actually in charge of staff or fighting units. This reflects the situation in the Republican army, where many officers were in theory available but mistrusted, even if they had not risen in rebellion. Several high-ranking men in the air force were shunted into non-active posts to do with aircraft manufacture, pilot instruction and negotiating in France or elsewhere to buy aircraft, while others, Hidalgo de Cisneros, Antonio Camacho, Carlos Núñez-Mazas and Antonio Martín-Luna Lersundi, possibly more in sympathy with the Spanish Communist Party's view of how the war should be fought, though the first was the only card-carrying member of the Party, occupied the Undersecretariat for Air, and held active command.[27] The following somewhat confused extract from a report of the Spanish Communist Party, dated 11 August 1938, underlines the point:

> Another problem in the Air Force is that of senior officers who are not employed or badly employed. There is a large number of officers such as [Gómez] Spencer, Camacho, Cascón and many others who are very badly used, and on the contrary, there are many majors and lieutenant-colonels who never fly. In the enemy air force, there are squadron commanders who are majors and commanders of larger units who are lieutenant-colonels and colonels. In our air force there is one major who commands a group and a few captains and lieutenants who lead wings and squadrons. This means that there are many officers of field rank for whom there are not enough posts, so they are in jobs where they do not do very much. One major with pilot's qualifications is head of transport in ports. Others are in command of aerodromes and supply organisations. In general the disorganisation (*desbarajuste*) is great.[28]

Colonel Camacho, now Undersecretary for Air, energetically set himself to the task of reconstructing the air force. The shortage of pilots was acute. Consequently, as early as 20 August 1936 applications to be trained as pilots, apprentice mechanics and specialists such as photograph interpreters and radio operators were invited. The main pilots' and mechanics' school was at Alcalá de Henares. Major Alejandro Gómez Spencer, the only senior officer to remain loyal in that airfield, was put in command. In early September, applications were invited for a course for observers. However, organization of schools for pilots began formally halfway through September 1936 with the first invitation to apply for a course at San Javier near Murcia. The naval air base here had declared itself in rebellion but had been crushed by forces from the neighbouring air force base at Los Alcázares, where almost all the officers, headed by Major Ortiz, had been loyal to the Republic. Courses were announced also for machine gunners and bomb aimers, and for aircraft mechanics. In December 1936 the Fighter School was set up under Captain (now Major) Sampil,[29] who previously had commanded the Bloch bombers acquired in France. For the moment, these posts were held by air force officers. In those early weeks of the civil war, as individual fighter planes duelled in the clear skies over militia forces on one side and the Legion, Moors, *Guardia Civil* and *Falange* and Carlist militias on the other, Ignacio Hidalgo de Cisneros, on his way to joining the Spanish Communist Party, strove to maintain the cohesion and loyalty of air force personnel. Unlike the insurgents, the Republican air force was headless because of the loss of the overall commander, General Miguel Núñez de Prado, and there was insufficient information about who could be trusted, accompanied by a justified distrust of most of the air officers in the major Republican bases of Getafe and Cuatro Vientos. Of forty officers in the *Oficina de Mando*, equivalent to the main air staff, only twenty-four were more or less loyal to the Republic. When, at the end of the war, Colonel Manuel Cascón was court-martialled by the victors he mentioned the names of senior air force officers, among them José Warleta, whom Colonel Pastor had sent to Paris for his own safety, and a number of others who were subject to constant suspicion because their personal opinions were known to be conservative and they were practising Catholics, who were likely to be unsympathetic to the Republic.

It is hard to form a coherent picture of the organization of the Republican air force over the next two years. When Largo Caballero's government fell in May 1937, it was replaced by the administration of Juan Negrín. Indalecio Prieto was retained and his role expanded to Minister of National Defence, with the air force, now termed *Fuerzas Aéreas de la República Española*, under an Undersecretary for Air (Antonio Camacho followed by Carlos Núñez- Mazas). The Republic, short of pilots, observers, gunners and other trained personnel from the start, made a vast effort to train men in airfields in the province of Murcia, in the extreme south-east of Spain, among them El Carmolí for high-speed fighter training, Totana for multi-engined aircraft and Los Alcázares for observers, gunners and armourers. Up till April 1937, 160 officers, non-commissioned officers and ground echelon men had been trained.[30] This was insufficient, as were the training courses held in private airfields in France for about 250 pilots. Consequently, Yakov Smushkevich (alias 'General Douglas'), the chief Soviet air force adviser, suggested training Spanish pilots in the Soviet Union.[31]

Three courses for training Spaniards were held in the USSR. On 14 January 1937, an expedition of 191 aspiring pilots under Major Manuel Cascón went to the USSR and received several months' instruction at the 20th Military School for Pilots at Kirobavad, returning in May 1937. On 25 May 1937 a second party of 118 trainees under Andrés García Lacalle, who was exhausted by the demands of commanding a squadron of fighters and required medical treatment, left for

Kirobavad, while a third group of about 50 left on 8 September 1937. Thus, some five hundred new pilots were trained to fly the *Chatos*, the *Moscas* and the *Katiuskas* and *Natachas* supplied by the Soviet Union. The first – and the best – Soviet pilots began to leave Spain in April 1937, replaced by aviators whose performance was generally judged to be of a lesser standard. By late 1937 most of the 772 Soviet pilots who served in Spain had left and the aircraft were being flown by men of little combat experience, unable to give cohesion to the squadrons. Consequently, veteran pilots and senior officers were constantly being relieved, thus depriving the squadrons of continuity of command. The few experienced squadron leaders, quite a number of whom had been killed at the outbreak of the civil war, disappeared from active service.[32] The last cohort of 185 aspirants was in the USSR when the Spanish war came to an end in March 1939. Of them, 104 of them elected to remain, and many flew for the Soviet Union against the Nazis; the others preferred to seek asylum in Mexico.[33]

While a few Republicans were trained in private flying schools in France, the largest number underwent training in the Soviet Union. The experience had a marked effect on the Spanish officers who led the expeditions.

Major Manuel Cascón, who in July 1936 had commanded a fighter wing at Getafe and had led the Republican air force in the Northern Zone, received orders to go to the base at La Ribera in Murcia and take charge of 191 young men who had been selected for training in the USSR. Accompanied by four other air force officers, the group sailed from Cartagena on 14 January 1937 on the steamship *Ciudad de Cádiz*. Landing at Feodosia in the Crimea, they travelled by rail to Kirovabad in Azerbaijan. According to the memories of one trainee, the Spaniards had to level the runway at Kirobavad themselves. Theory classes were given in the packing cases in which the training machines were brought. After preliminary training, they moved to the R-5 and then to fighters or other advanced training. The courses were very intensive and only one day in ten was allowed for rest. Trainees remembered Major Cascón as highly active, severe and a top-class gymnast in the exercises that the Russians insisted on. He was very concerned that the trainees should be seen as 'Spanish officers'. The Russians should see this clearly, he would stress (*que se note-e-e* or 'make it obvious!'), recalled one trainee. Major Cascón wrote a report which has not been found, but there is evidence to suggest that he was shocked by the purges at the time of senior Russian officers and by the 'suffocating propaganda' of the Soviet system. In July 1937, the expedition returned to Spain. Cascón became Inspector of Flying Schools and commander of the 2nd Air Region.[34]

Ex-sergeant pilot Andrés García Lacalle, who had played such an important part in the early months of the war, was in serious need of medical treatment by late May 1937, when he was awaiting his journey to a Russian sanatorium. While Lacalle was waiting for his orders, he helped to select the men for the second expedition to the USSR. It was here, at the Los Alcázares base in Murcia, on 29

May, that he learned about one of the most serious bombings of neutral ships in the Spanish Civil War, and one which came close to provoking a major European conflict. On that day the overall Republican air force commander, Colonel Ignacio Hidalgo de Cisneros, was at Los Alcázares, together with a number of Russian pilots of SB-2 *Katiuska* bombers. Excitedly, the Russians were looking forward to bombing the brand-new Nationalist cruiser *Canarias*, reported as at anchor at Ibiza, the Balearic island under Francoist control. Within a few hours two *Katiuskas* landed, excitedly waggling their wings. As the crews left the planes, they shouted, as Lacalle heard, *Bombili Canarias*! Lacalle began to interrogate one of the Russians. How did the observer know it was the *Canarias*? The Russian said that he recognized the outline and that his plane had been fired on from the ship. One of the *Katiuskas* had failed to hit the cruiser but the other had scored impacts near the funnel and towards the bows. Early the next day, Hidalgo de Cisneros and Lacalle drove urgently northwards along the coast to the Ministry of National Defence in the wartime Republican capital of Valencia. On the way, Hidalgo de Cisneros told Lacalle that the German fleet had shelled the Mediterranean port of Almería.

García Lacalle's version, however, leaves much out.[35] The cruiser was not the Spanish Nationalist *Canarias* but the German pocket battleship *Deutschland*. This warship was participating in the international patrol which was intended to stop merchant ships carrying war material to Spain illegally in terms of the general European agreement not to supply either side. However, the German naval authorities ought to have notified the Spanish government that the *Deutschland* was in harbour at Ibiza, given that its usual berth for rest and refuelling was Palma on Majorca.

The *Katiuskas* which bombed the *Deutschland* were part of a shipment which had arrived at Cartagena at the beginning of May. The circumstances, as later investigated, were that Palma harbour was being regularly bombed by Republican aircraft. For its own protection, the *Deutschland* had dropped anchor at Ibiza. British, French, Italian and German warships were properly exercising their functions in patrolling on the high seas to prevent arms reaching Spain, but not while they were in Spanish territorial waters, where they were visitors and, in the case of German and Italian warships, open enemies of the Spanish Republic, which they no longer recognized, and thus appropriate targets.

At about 7 pm local time, two *Katiuska* bombers from Los Alcázares swept out of the setting sun over Ibiza. They dropped bombs on the *Deutschland* from about 3,000 feet. A large part of the crew of the German pocket battleship was on the mess deck. Thirty-one German sailors were killed and seventy wounded. The ship made full speed to Gibraltar to bury the dead, whose remains were later repatriated to Germany, and to hospitalize the most seriously wounded.

There are contradictions in identifying the Russians who bombed the *Deutschland* but a reliable account identifies them as the pilot M. G. Khovanskii and the navigator/bomb aimer Georgii K. Levinskii.[36] While the pilot insisted that he could tell a German pocket battleship from a Spanish cruiser, Captain Kuznetsov, the Soviet naval attaché in Republican Spain, was doubtful. If the German commander opened fire before his ship was bombed, as the Soviet pilots claimed, his reaction would seem to have been justified if two bombers came in low and fast from what for him was enemy territory. Perhaps he did open fire first, but it was in the interest of the Spanish Republican government to insist that this was the case even if it were not so. Nor could the Spanish government reveal that Soviet airmen had dropped the bombs, because the presence of Russian aircrew was officially secret even if Russian aircraft were not. The entire matter was embarrassing because the Republican government would have to admit that Spanish airmen were incapable of identifying a Spanish warship from a low height.

Actually, the Republican government was let off the hook by the intemperate German reaction. It took a considerable time for Hitler's staff to calm down his rage. For the Fuehrer, the Spanish Republic, with which Germany had had amicable relations until the war, was a pirate without normal rights. He wanted to have the German navy bombard Cartagena or Valencia and to declare war openly on the Spanish Republic. These two ports were well-armed with long-range coastal artillery, so the German admirals persuaded Hitler to order a demonstration shelling of Almería, a smaller port further south, which had only short-range guns. In the early morning of 31 May, four German destroyers and the battleship *Admiral Scheer* were observed contemptuously by a British destroyer as they fired 275 shells into the undefended port of Almería, killing nineteen people and wounding fifty-five. Indalecio Prieto, the Spanish Minister of National Defence, wanted to launch an outright attack on German warships in Spanish waters, which might well have provoked a major conflict, but was dissuaded by his government, and in particular by the communist ministers, for whom such a result would have been a complete disaster for Soviet foreign policy.[37]

Soon after the *Deutschland* episode, Andrés García Lacalle went to the USSR for medical treatment at Kislovoch in the Caucasus. Following this, he took charge of the course for training Spanish pilots at Kirovabad. This course was for about two hundred fighter and bomber personnel. Lacalle reports that the course was demanding and monotonous. The trainees were not allowed to go into the local town, except when some visiting actors came to the theatre of Kirovabad. The Spaniards had to march through the town, looking neither left nor right, when they went for their meals. As in the Republican forces, where commissars had been appointed principally to ease relations between the men and their often distrusted officers, the trainees had a commissar, a Russian called Mirov, who was a Latin American expert and thus spoke Spanish as he gave daily lectures during mealtimes, as though the trainees were in a monastery. On the other hand, Lacalle was most impressed by the seriousness and constancy of the Russian instructors. The extreme efforts of the instructors with the weakest student pilots meant that almost no student was failed. Some instructors spent as long as six hours a day at the dual controls of I-16 fighters. Lacalle recalled having, as Spanish commander, broken the rules, gone out and bought vodka and beer for the exhausted instructors, one of whom, however, sneaked to the commissar. When Lacalle told the other instructors who the sneak was, they replied that they could do nothing to him. Just like Major Cascón in the first expedition, who had been appalled by the level of political correctness imposed on the Russians, so Lacalle was saddened that the instructors, whom he describes as 'magnificent and virile men', had to thank the sneak for reporting them to the commissar. One or two other trainees clashed with the commissar and needed the help of the Spanish ambassador in Moscow to leave the USSR at the same time as the other students. Fortunately, the commissar knew that his fate depended on the successful completion of the course by all the trainees, which meant that he had to modify his political rigour to some extent. In fact, save for a few losses due to flying accidents, and one absolute failure, the entire expedition returned to Spain in February 1938, where they replaced Soviet pilots who had by now been repatriated.[38]

The tone of the reorganizations of the air force that were regularly published by the Republican Undersecretariat for Air read wordily and as if they were being set up for a country at peace rather than one fighting a desperate war. Republican administrators were concerned with seniority lists, how naval flyers were to be incorporated with military flyers into one single Army of the Air (*Ejército del Aire*) and details of promotion and retirement policies. This contrasted markedly with the tone of organization on the Nationalist side. This was very much in military style, brief and to the point, which was how to win the war.

As for training of new Nationalist aircrew, in the spring of 1937, Italy had provided teaching personnel for a school for mechanics and other specialists in Málaga. Altogether, some five hundred Nationalist pilots were trained in Spain, to be added to 140 in Italy and Germany. In December the first 106 Spaniards would arrive for flying instruction in Italy, in schools to supplement the Nationalist centres at Logroño, Cáceres and Seville. Thus the number of Spaniards trained on either side was roughly comparable, but the large number of professional German and Italian aircrew personnel compared with the ever-falling numbers of Russians had a major effect on the competence of the Republic's air force.[39]

Pre-war Spain had had three air force groups, called *Escuadras*, in Madrid, Seville and Barcelona. Republican Spain was now divided into eight air districts (*Regiones Aéreas*), reduced to seven when the North fell to Franco's forces, each one of which was split into sectors. Some districts had several wings, usually with three squadrons each. While the Nationalists had only about one hundred aerodromes in 1937, the Republican air force had a total of some four hundred, some permanent, some semi-permanent, some occasional and some to be used in emergencies only, but all listed and with their establishment of ground troops laid down.[40]

Figures for the numbers of aircraft tend to vary. The main point is that the Soviets did not substantially increase their contribution, nor did they send newer models in the way that the Condor Legion did. Furthermore, the general view was that the later Russian crews were not up to the standard of the earlier ones and the Spanish airmen trained in the USSR were not of the quality of the professional and highly trained and regularly replaced Germans of the Condor Legion and Italians of the *Aviazione Legionaria*. Many Russian pilots had to learn basic combat techniques in Spanish skies. In addition, Soviet pilots sometimes landed at the wrong airfields, and even in enemy ones. In December 1936 alone, fourteen Soviet aircraft were lost by pilot error.[41]

Segovia, Brunete and Belchite

In the summer of 1937, the new Republican government, formed on 18 May under Juan Negrín, in which Indalecio Prieto was now completely in charge of the war as Minister of National Defence, with Colonel Vicente Rojo as his chief of staff, launched a number of operations to try to take the pressure off the northern front and to relieve Nationalist pressure on Madrid.

The first of these was northward from Madrid as far as the provincial capital of Segovia. It is this battle which Ernest Hemingway uses as the background for his famous novel of the Spanish Civil War, *For Whom the Bell Tolls*. The attack began on 30 May with support from *Katiuska* bombers, *Rasante* and *Natacha* low-level bombers and thirty-two I-15 *Chatos* and I-16 *Moscas*.

Republican ground troops were unable to achieve their aims and the Nationalists were soon in control of the skies. Every report from the ground commanders, including one from the Soviet adviser General Maximov, puts the blame firmly on the shoulders of the air force. Aircraft arrived late, remained for a very short time over the battlefield and bombed from too high a altitude.[42] Possibly this was because the Soviet pilots who had been in Republican Spain since the previous autumn were exhausted and about to go home or, on the other hand, that the Russians who had newly arrived were not too skilled at handling their machines in Spanish conditions. In contrast, the Nationalist ace and profoundly experienced squadron commander, Joaquín García Morato, used the technique of maintaining two aircraft permanently on watch over the battle area so that Nationalist aircraft would be able to reach the appropriate height to attack their enemy. Remaining over the front until fuel was almost exhausted was a technique used by Nationalist pilots throughout the war. It was chancy and not practised by the Republican pilots who, unlike their enemy, had to cross the Guadarrama mountain range to return to their bases.[43]

The second of these land battles, in which the new Popular Army of the Republic was trying out its newly developed and rather shaky skills, was that of Brunete, a town some 15 miles west of Madrid, where the Republican army aimed to try to break the Nationalist lines.[44] The Republic's elite troops, though not of the military quality of Franco's Moors and legionaries, backed up by aircraft, struck on the morning of 6 July, in extreme temperatures where thirst was a major problem. Pilots would find that, at 44 degrees centigrade on the ground, the water-cooling systems in aircraft did not function until their planes reached a height of perhaps 6,000 feet.

The area which was to be attacked was protected by a wing of Fiat CR-32 *Chirri* fighter biplanes and a wing of thirty-one Romeo-37 *bis* up-to-date Italian two-seater reconnaissance biplanes with two machine guns firing forward and another in the observer's cockpit, as well as some light bombs under the wings. This was a more powerfully engined version of the Romeo-37 *bis* which had arrived in Nationalist Spain in January and February 1937, one squadron of which was formed into a wing commanded by Juan-Antonio Ansaldo, the airman who had been badly burned when his Puss Moth had crashed in Portugal in the accident in which General Sanjurjo, the putative leader of the military uprising, had died.

It was at this time that the Italians decided to adopt the ground-strafing technique known as the 'chain' or cadena, which Spanish pilots of the German He-51s had invented. In the chain, each aircraft dived on its target with the engine idling. As it pulled out, the pilot opened the throttle to climb away at top speed. Unfortunately, the Romeo engines tended to cut out at this tense moment, so pilots were dissuaded from using the *Cadena* tactic.

In the Spanish Civil War, Republican army units in general, commanded by inexperienced and insufficiently trained officers, were not capable of exploiting their initial success. Moreover, the well-organized resistance of small insurgent garrisons tended to hold up the Republican advance and, most importantly, Franco's logistics were of high quality. At Brunete, Junkers-52s and He-45 two-seater, light bombers and aircraft of the *Aufklärungsgruppe* or reconnaissance wing of the Condor Legion, known as 'Turkeys' or *Pavos*, and now commanded by the very experienced Captain Cipriano Rodríguez, a world champion in the circuit races of the 1920s, together with a squadron of Aero-101s, the Czech light bombers sold to the Republic but captured at sea by the Nationalists, were flying south from their bases in northern Spain and bombing and igniting the dry woodlands in the battle area within a few hours of the Republican advance. By 9 July, Lieutenant Colonel Lecea, head of the Nationalist Central air command, had assembled 150 machines at aerodromes within reach of the battlefield. The Nationalists could launch Savoia SM-79 bombers, Dornier Do-17s, Bf-109 fighters, He-111 bombers and Italian CR-32 fighters, as well as the older Ju-52 bombers, now flown by Spaniards.

On 7 July 1937, the Italian 19th and 20th squadrons participated in four separate aerial battles near Brunete, engaging Republican formations twice in the morning and twice in the afternoon. They claimed three light bombers, probably *Natachas*, one *Katiuska*, seven *Chatos* and five *Moscas*. This was how Captain Enrico Degli Inverti, commanding officer of the 19th squadron recalled his victories that day.

> Each one of us chose his quarry and the melée began. Our guns splendidly spat out a barrage, and our adversaries replied in kind. It was a matter of life or death. I pounced on a *Rata* [I-16] and shot at it. It appeared that I had scored a direct hit. I kept following him until I thought that he was clearly falling away. However, as I broke off my chase he zigzagged, dropped a little further and then climbed. He attempted to turn onto my tail, so I quickly hit him again. He pulled up abruptly after diving down a few hundred feet, so I fired at him once more. It looked to me as if the bullets had found their mark – the tracers clearly indicated that I was aiming correctly – but the *Rata* pilot continued to defend himself.
>
> I persevered with my foe, despite now feeling that I was possibly coming under attack. I looked over my shoulder and spotted three enemy aeroplanes, still at a distance, heading in my direction with their guns blazing. Moments later my prey finally fell headlong into a thickly wooded area. Staying with him had made me lose precious height, and as I looked up I could see that the fighting was still continuing above me.[45]

The pilot who had been shot down by Captain Degli Inverti was almost certainly Lieutenant Aleksey Sergeyevich Trusov, who had been in Spain for just a matter of weeks. Another damaged I-16 *Mosca* crash-landed in Nationalist territory and its pilot, Lieutenant Khoziainov, was taken prisoner.[46] Fortunately for them, the Nationalists had orders not to mistreat captured Soviet pilots, who were useful for exchanging with their Francoist opposite numbers.

No sooner had the 2nd fighter squadron of the Condor Legion (2 J/88) reached Ávila after flying that morning from Burgos than all nine of its Bf-109s were refuelled and sent off again as escorts for bombers attacking designated targets.

On 7 July 1937 dogfights took place over Madrid between Fiat CR-32s and Soviet squadrons led by Ivan Eremenko, a newly arrived Soviet pilot, who would in September become overall commander of the Soviet fighter group, and by Ivan Lakeev and Alexandr Minaev, who would be shot down and killed over Brunete on 15 July.[47]

Three I-15s from the 1st squadron were lost; Flight leader Serov and Yakushin managed to nurse their badly damaged biplanes back to base. Nikolai D'yakonov, who was leading an I-16 flight, suffered serious wounds in combat and died later that day after landing in Republican territory.

After this confusing battle with its contradictory reports, it seems that the Republican pilots claimed five CR-32s at least while losing three I-15s and suffering several damaged. The *Aviazione Legionaria* claimed seven victories.

One of the Soviet squadrons, piloted by skilled flyers such as Mikhail Yakushin, Anatoli Serov and Yevgueni Stepanov, drawn from the 1st squadron of *Chatos*, specialized in night flying. Yakushin was the first pilot ever to shoot down an aircraft at night, a feat which he achieved when he downed a Ju-52 over Brunete on the night of 25–26 July 1937, for which he was awarded the Order of the Red Banner. This was his account

> At midnight we received a telephone report of an enemy bombing raid on Republican troops near El Escorial. It was the first time that we had approached the front after dark. The search area was outlined by the fire started by the bombing. Serov remained at our initial altitude of 6,500 ft, while I climbed 3,250 ft higher. My luck was in, for ten minutes later I spotted an enemy bomber heading towards me. He would not get away.
>
> Having let him pass, I turned and began to approach him at the same height from his right and behind. We had learned by then that the Junker's fuel tank was positioned near where the right wing joins the fuselage. Having approached the target and slowed down, I fired at that area. Flame appeared along the right side of the bomber's fuselage. Almost at once the enemy gunner responded, but he was too late. His bomber was already going down in flames. I followed him down almost to the ground.[48]

Serov shot down another enemy plane the following night. The *Chatos* of his squadron were equipped with flame dampers on their exhausts and flash guards on their gun muzzles. Their technique was to spot an enemy bomber as it passed through searchlight beams and then try to intercept it by homing on to its exhaust flames.[49] Yakushin returned to the USSR in November 1937 and went on to enjoy a distinguished career.

On 6 and 7 July, a *Mosca* and a *Katiuska* were downed. On 8 July, another *Mosca* was hit by anti-aircraft fire, and several *Chirris*, He-51s and other Nationalist planes were lost. On 16 and 17 July, Nationalist bombers were attacking Republican aerodromes, and on 18 July, the anniversary of the Glorious National Movement as the military uprising was now called, they downed twenty-one Republican planes.[50]

The battle of Brunete ended with a small advance for the Republican forces. The Popular Army was as yet not able to carry out successful attacks. As for the air force, the Republican professional army commanders complained that the air force, unlike what was known to be the case on the Francoist side, was not liaising properly with the land troops. A report from the XVIII Corps of the Republican army reads tellingly:

> Our men suffer the fire of enemy aircraft, which are flying all the time over our lines. This tactic depresses the morale of our forces. The enemy uses few machines, but continuously, so our troops feel that they are constantly being watched, while our air force does nothing to counter this impression. The enemy's constant machine-gunning paralyses our will to fight.[51]

One way of interpreting the lack of air support of which the Republican army commanders complained may perhaps be the unusually high number of accidents suffered by the Russian pilots. Compared with a Nationalist figure of thirteen aircraft lost between October 1936 and 1 July 1938 by forced landings and accidents, the Republican score was an appalling 147.[52]

The American mercenary pilot Frank Tinker flew at Brunete, assigned together with his fellow American Albert Baumler to the squadron of I-16 *Moscas* commanded by Ivan Lakeev. The latter had been in Spain since the previous November and would be replaced in August 1937 by new Russian pilots. Returning to the USSR, he would rise to the rank of major general. Interestingly, Tinker recalled that the new Russians appeared substantially younger than the veterans he had known, and, after flying with them, he judged the newcomers to be less capable than *los rusos de antiguo* or 'the old Russians', as the ground staff called them.[53] His fears were justified when one pilot destroyed his landing gear in a hard touchdown and another overshot the runway on which they landed on their way from Los Alcázares to Barajas. Finally, Tinker's *Moscas* were sent to a new airfield at Manzanares el Real, some 50 kilometres north of Madrid. Here, on a high (12,000 feet) aerodrome with thin air, the incapacity of the new pilots was shown up even more. They did not realize that in the thin air of the Sierra, they needed more speed and runway.

A greater challenge faced the new pilots, for on 12 July Messerschmitt Bf-109 fighters, not seen before in central Spain, appeared. Unlike the slow He-51, the Bf-109 was a fearsome foe. And when Dornier Do-17 bombers were spotted, the Soviet fighter pilots found that the Germans flew too high and fast to be caught.

In April 1937 a number of Italian *Chirri* fighters had been transferred to the Spanish Nationalist air force. Here they joined the five transferred four months earlier to form the basis of the first Spanish *grupo* ('wing' in RAF terms) equipped with Fiat fighters. Grupo 2-G-3 consisted of thirteen aircraft and fifteen pilots, which were divided into two squadrons of six fighters each. The thirteenth CR-32 was the personal machine flown by the group commander, Joaquín García Morato. On 17 May, García Morato had been decorated with the highest Spanish medal, the St. Ferdinand Cross with laurels (*Cruz Laureada de San Fernando*) for his actions up until 18 February 1937. Until the latter date, he had made 150 sorties and participated in 46 combats, claiming 18 victories.

By the summer of 1937, the Italian air force or *Regia Aeronautica* had spent enormous sums in Spain. These sums would be tripled by the end of the war.[54] At the end of July 1937, the Italian government agreed to provide fifty-two more Fiat CR-32 fighters, ten Ro-37 low-level bombers and twenty-four ultra-modern Savoia S-79s. At the end of 1937, a further twenty-two S-79s arrived, allowing for four bombing wings to be formed. The arrival of more Fiat CR-32 fighters allowed García Morato's wing to add another squadron.

One of the two Spanish Fiat fighter squadrons was commanded by Julio Salvador and the other by Angel Salas, both aces in the Spanish Civil War. Two of the pilots had previously served in García Morato's 'Blue Patrol' or *Patrulla Azul*, while the remaining twelve were chosen according to their experience in fighters. These two squadrons, whose planes were undergoing repair at Tablada, were hurriedly made ready.

Very soon after the initial Republican attack at Brunete and thanks to the organization set up by General Mario Bernasconi, who had replaced Colonel Velardi as commander of the *Aviazione Legionaria* in March 1937, seventeen Fiat CR-32 *Chirris* had flown from Vitoria in Navarre to the beautiful and austere walled city of Avila, 100 kilometres from Madrid but much higher than the Spanish capital. From Avila, Brunete was a mere few minutes flying time. So, within about four days of the start of the battle of Brunete, thirty-one Italian CR-32s, and seven more under Spanish command, had reached the area of fighting.

When the challenge to the Republican air force over Brunete is looked at closely, the complaints from the army staff about the absence of aircraft become understandable. Republican aircraft were being fiercely challenged all over the central area of Spain. On 12 July, for example, a huge dogfight took place near El Escorial, the great monastery and pantheon of the kings of Spain north of Madrid. Then twenty-nine I-16s from the squadrons commanded by Ivan Lakeev, Nikolay Vinogradov and Petr Shevtsov, together with eight I-15s of Captain Ivan Eremenko's squadron, attacked a group of forty He-51s and CR-32s. Both sides initially claimed nine victories each, although both admitted losing only one fighter.

A few hours later, a squadron of Russian R-Z reconnaissance bombers was attacked by what the surviving crews described as 'high-speed monoplanes' (probably Bf-109Bs, although they were not yet identified as such). I-16 *Mosca* fighters escorting the bombers attempted to engage the aircraft in what was probably the first battle between the I-16 and the Bf-109, the two most advanced Russian and German fighters of their day. The I-16s suffered no losses, but the Bf-109B of *Unteroffizier* Guido Höness, who had shot down two R-Zs before he was killed, was downed. The victory was attributed to a number of Soviet pilots, but it is possible that it was achieved by Petr Butrym, another recently arrived Soviet pilot.

Once the Republican attack at Brunete had become bogged down and as the Nationalists counter-attacked, they were supported by Bf-109Bs of the Legion Condor, which undertook ground-strafing missions in the wake of the He-111 bombers. Captain Gotthard Handrick, commanding the German fighters, reported triumphantly:

> The success of our low-level attacks was quite considerable. Spanish observers reported that on one occasion following our low-level attacks on a 150 m. stretch of foxholes, 100 Red Spaniards had been taken care of by our machine guns. It was possible to make this determination because immediately after our attack the foxholes were taken by Spanish Nationalist infantry.[55]

One of the most vivid accounts of the air war, added to a careful considerations of many issues, was written by Lieutenant Günther Lützow, a squadron commander in the Condor Legion, fighting over Brunete and referring to 21 July 1937:

> The battles on the Brunete Front had reached their zenith. Despite our air superiority, the Red bombers attacked again and again [...] a small group of volunteer fighters with limited means [was fighting] against a tremendous superiority in personnel and materiel. What made us different from our opponent was the quality of our pilots and our aircraft.[56]

Lützow, as a fighter commander, would have been thinking of the Messerschmitt Bf-109, which was also the subject of conversation among the Russian and the American pilots of the Republican fighter squadrons. The latter were most impressed by the Bf-109's three streams of bullets. It had two machine guns and another which fired through the propeller hub and which the Republican pilots suspected was a 20mm or even a 37mm cannon, although it was in fact only a third 7.7mm machine gun. But this gave it 50 per cent more weight of bullets than the I-16.

The clash between the I-16 and the Bf-109, both monoplanes with enclosed cockpits, retractable undercarriages and very high top speeds (the latest Bf-109 could by now reach 290 mph), marked a change in the tactics of air fighting. Fighter battles now became less like swarms of gnats and more like vast, sprawling affairs over hundreds of miles, as they would be in the Second World War and would remain until the first battles between jet fighters in the Korean War of 1950–2.[57] In Spain, however, where the Republican formations were heavy, consisting at times of forty or fifty fighters, the Nationalists would use a minimum of perhaps two or three manoeuvrable and agile squadrons of *Chirris* or Condor Legion Bf-109s, which with much smaller numbers could keep the battlefield under watch and the Republican ground troops in a constant state of tension.

The introduction by the Condor Legion of the fast Dornier Do-17 bomber, as the American fighter pilot Frank Tinker noted, meant that unless the Republican fighters had previous notice, by the time they could reach an appropriate height to attack, the German twin-engined 'Flying Pencil' was well on the way home to its base at Avila 100 kilometres from the Spanish capital.[58] Likewise, now that the lumbering Ju-52s were kept out of the way of the Soviet fighters, it was no longer a simple matter for Republican fighter aircraft to tackle a German He-111 or an Italian Savoia-Marchetti S-79 (much faster than the Savoia-Marchetti S-81s supplied to Franco in 1936). Furthermore, these new bombers flew to their targets at heights of up to 15,000 feet, often not being even spotted by observers, and then made their final approach in a practically silent power glide, opening up their throttles only for the last mile of the bombing run. After this, they sped off at minimum height and maximum power.[59] By the time Republican fighters were in the air, the bombers had dropped their bombs and were flying at speeds not too far short of what Republican fighters could reach. They were practically invulnerable.

Chapter 8

The Spanish War moves to the east – Aragon, Belchite, Teruel: Franco splits the Republic in two

The Spanish war now moved into eastern Spain, principally Catalonia and Aragon. Here, the major pre-war air base was at El Prat, today the civil airport of Barcelona, then headquarters of the air force's 3rd *Escuadra* or Group, commanded by Lieutenant Colonel Felipe Díaz Sandino, who from the first moment took a resolute stand against the insurgents by dispatching aircraft to bomb the rebellious barracks of the city and force the surrender of the naval air base. According to the results of the post-war court martials, of twenty-nine officers at El Prat, only sixteen remained loyal to the Republic.[1]

In Barcelona the anarcho-syndicalist workers' union, the *Confederación Nacional del Trabajo* or CNT, also played an important part, together with the militarized *Guardia Civil*, in the repression of the military insurrection. Lluis Companys, President of the autonomous government of Catalonia, the *Generalitat*, offered to hand over power to the CNT leaders but was asked to stay on, while the Anti-Fascist Militias Committee which took over laboured to combine a social revolution with a military campaign. For several weeks Barcelona was a city where, in the British writer George Orwell's words at the beginning of his book about the Spanish war, *Homage to Catalonia*, 'the working-class was in the saddle'.

In Aragon, to the west of Catalonia, the insurgent forces occupied a line running from Huesca in the extreme north, down to Zaragoza and then to the salient of Teruel. Their organization, mobile reserves and communications were efficient enough to resist the attacks of an enemy which was virtually bereft of military structure and immersed in revolutionary chaos. Several columns of militia, imbued with anarchist concepts of hostility to military order and discipline, as well as to the attempts of the Madrid government to create a proper army, left Barcelona in columns of buses and requisitioned cars, with the aim of taking Zaragoza, but never did so.[2] For months there was dissension in the Anti-Fascist Militias Committee between the CNT, the Unified Socialist Party (*Partit Socialista Unificat de Catalunya*), dominated by the communists, and the revolutionary and anti-Stalinist Marxists or *Partido Obrero de Unificación Marxista* (POUM), in whose militia George Orwell fought. It took until 24 October 1936 to dissolve the Anti-Fascist Militias Committee and to bring the militias and their parties under the control of the *Generalitat*, the autonomous government of Catalonia. Yet, since the *Generalitat* did not have responsibility for defence, its own assumption of military powers was unconstitutional.

As the militia columns advanced towards Zaragoza, the political void was filled by the creation of a body called the Council of Aragon (*Consejo de Aragón*) which presided over a widespread social and anti-clerical revolution, collectivizing land and burning churches. The chaotic situation continued for close on a year until, as a result of the internal struggle which took place in Barcelona from 3 to 6 May 1937 between the Republican government on the one hand and the revolutionary CNT and POUM on the other, Largo Caballero was forced to resign as Spanish Prime Minister, to be replaced by his fellow socialist Juan Negrín. During the summer of 1937, Negrín's new Spanish government energetically suppressed the revolutionary POUM, abolished the Council of Aragon and took over control of public order and military organization in Aragon and Catalonia. Anarchist ministers were dropped and the two military ministries (War and Navy and Air) were merged into one Ministry of National Defence, still under the leadership of Indalecio Prieto, and with Colonel Antonio Camacho still retaining the post of Undersecretary for Air. The 'Army of Catalonia', which had been formed from militia columns, became the new Army of the East, under General Pozas. During all this time, the failure to organize a properly structured army in Aragon and Catalonia had allowed the insurgents to build up their strength.

As for aircraft, on 20 July 1936, the insurgents who had taken over the base at Logroño sent some Bréguet-19 fighters to help defend Zaragoza against the advancing militia.[3] Later the insurgents at Zaragoza were supplied with twenty-four German He-46 reconnaissance machines with bombing capacities.[4] On the Republican side, Lieutenant Colonel Díaz Sandino became overall commander of the local air force and Counsellor for Defence of the autonomous *Generalitat* of Catalonia. An advanced aerodrome was established at Sariñena, with a squadron named 'Red Wings' or *Alas Rojas*, which included one Bréguet X1X, some Nieuport-51 fighters and a couple of other aircraft. The appropriate strategy – and one which the insurgents would probably have adopted had they been in that situation – would have been to concentrate every aircraft they had – as Franco had done on 5 August 1936 to protect his military transports from the Republican fleet in the Strait of Gibraltar – and to bomb all the military installations in Zaragoza while the insurgents were still establishing their authority.

In Barcelona, the CNT promoted a flying school with a few machines which had been confiscated from their owners. Somewhat optimistically, the aim was to train ten pilots per month on the basis of fifteen hours under dual control and ten solo. For technical aid they counted on the support of Major Miguel Ramírez de Cartagena, who was second in command at El Prat base. However, the anarchist enterprise collided with the interests and objectives of the *Generalitat*, which wanted to form its own training school, and with the newly created Ministry of the Navy and Air which had come under the socialist Indalecio Prieto from 5 September 1936. Colonel Díaz Sandino was making his own selections of men to train at El Prat as well as at the Republican government's school at San Javier in south-east Spain. So, after three months of fruitless argument, the CNT had to accept that the training of pilots was the responsibility of the government of the Republic and not of a workers' union.

Belchite

Operations in Aragon in the first few weeks of the war were a reflection in miniature of those which were happening in the mountains to the north of Madrid. Individual insurgent airmen, among them the future ace Captain Angel Salas, cruised over the area attacking any enemy machine or militia column they could see. In his first month, Salas himself completed fifty missions and 116 flying hours and collapsed with exhaustion.[5] On the Republican side, occasionally, a bomber from Barcelona would fly over Zaragoza and drop bombs.

One of these raids contributed to the Francoist myth of divine protection. At about 2.30 am on Monday, 3 August 1936, a Republican aircraft dropped three bombs on Zaragoza, two of which fell inside the famous Cathedral known as El Pilar but neither exploded. Over following decades, the religious emotion aroused by what the Nationalists had begun to refer to as their 'Crusade' led to frequent references to the miracle of the unexploded bombs, though given the general disorder, neglect and indiscipline on the other side, it might perhaps have been miraculous if any of them had indeed exploded.

One of the first fruits of the re-establishment of central government authority in Aragon and Catalonia in the summer of 1937 and the creation of the new Republican Army of the East was the battle of Belchite, part of a series of operations directed against Zaragoza in order to try to draw off Francoist forces from Santander and Asturias. Beginning on 24 August 1937, rather late because this was only two days before the Italian *Corpo di Truppe Volontarie* took Santander, eighty thousand men of the elite communist-led divisions and International

Brigades of the Republican army launched an offensive towards Zaragoza, with some immediate success. However, the Republican forces committed tactical errors similar to those which had bogged down their July advance at Brunete, and by 29 August, the offensive was paralyzed. Then the Nationalists, relying on their well-fortified line to resist the initial Republican assault, and having brought up reinforcements, counter-attacked and threw back their enemy.

The air battles in this operation were heavy. The Nationalists had recently carried out a reorganization of their air force, dividing it into an *Aviación Independiente*, which came directly under the orders of Franco's headquarters, and included the Italian *Aviazione Legionaria*, the German Condor Legion and the Spanish-crewed Ju-52 bombers and Fiat fighters, and the *Aviación de Cooperación*, consisting of the reconnaissance and low-level Ro-37, He-51s and He-46s. These were Italian and German bombers flown by Spaniards and came under the direct orders of the large infantry ground units. The new arrangement was an attempt to solve the question, a matter of frequent dispute in many national armed forces at the time, of how to use an air force, that is, whether to divide it or to keep it under central command.[6]

By the afternoon of the Republican attack on 24 August, Major Andrea Zotti's wing of Fiat CR-32 *Chirris*, whose machines bore the insignia of the *Asso di Bastoni* or 'Ace of Clubs', had arrived under his command at the aerodrome named after General Sanjurjo close to Zaragoza. They were followed by García Morato's fighter wing and soon by the three squadrons of the *La Cucaracha* wing, as well as by S-79 bombers from their base at Soria. On 30 August, another Italian fighter wing arrived from the northern front. The Nationalists now had air superiority in Aragon. Under their protection, the He-46 low-level bombers began continually strafing enemy positions and road transport, as they attempted to relieve the Republican siege of the small town of Belchite. The Italian pilots felt profound admiration for their Spanish opposite numbers who spent their evenings carousing in Zaragoza and their days flying low and bombing enemy positions from slow Ju-52s and He-46s.[7]

On the Republican side, the serious losses of aircraft in the north since April and at Brunete in July had not been made up by shipments from the USSR. The Republic was able to put up only three squadrons of *Moscas*, four of *Chatos*, three of *Katiuska* bombers and a wing of *Natachas*.

Both sides lost aircraft in fierce battles. Finally, after desperate resistance, Belchite was occupied on 6 September 1937 by the Republican forces which had besieged it since 26 August. Franco decided not to try to retake the town and moved his forces back north to Asturias to complete the occupation of the northern Republican zone. After the war Belchite was rebuilt on another site, and the ruins have been preserved until now.

A Republican bulletin from the beginning of the battle reported that a captured Francoist airman, Major Joaquín Pérez Pardo, had a serious stomach wound and was in hospital.[8] He had commanded the wing of He-46 reconnaissance and low-level bombers stationed at Zaragoza. The commander of the Republican air force, Colonel Ignacio Hidalgo de Cisneros, visited him in hospital in Barcelona. In the Spanish Civil War, men who had been close comrades, even personal friends, found themselves on opposite sides. In this case, during the Moroccan wars of the 1920s, Pérez Pardo had flown in Hidalgo de Cisneros's plane as his observer. For the latter, it was an embarrassing moment to meet his former comrade, now his enemy, but he was able to relieve the wounded man's fears that he was to be shot out of hand. Pérez Pardo believed in his side's propaganda which proclaimed that the Republic had received immense reinforcements of Russian soldiers. Hidalgo de Cisneros gathered that there was panic in the Aragonese capital, Zaragoza, at the thought of the city being taken by the Republic. But no further advance towards that city took place. When the Republican air chief came to visit his ex-comrade once again, he learned that Pérez Pardo had died during the night.

Italian aircraft in Spain

Without counting aircraft being repaired or in reserve, which amounted to a third of the force, halfway through October 1937 the figures for the *Aviazione Legionaria* on the Spanish mainland, that is, without the Italian planes on Majorca, totalled as follows:

Fiat CR-32 *Chirri* fighters	108
Romeo-37 bis low-level bombers	20
Breda-65 low-level bombers	6
Savoia-Marchetti S.79 bombers	6
Savoia-Marchetti S.81 bombers	17
BR 20 *Cicogne* fast bombers	3
Total	160

In the Balearics, harassing Republican sea traffic, there were as follows:

S-79 bombers	11
S-81bombers	6
Fiat CR-32 fighters	10
Macchi M-41 fighter flying boats	2
CANT- Z-501 flying boats	2
Total	31[a]
[a] Figures from Pedriali, *Guerra di Spagna e aviazione italiana*, p. 281.	

Soviet aircraft manufactured in Spain

During August 1937, the Republican air force received the first three *Chatos* manufactured in Spain. The decentralization process of relocating the aircraft industry in Catalonia and eastern Spain had been begun by Prieto in September 1936 when he became Minister of Navy and Air. Yakov Smushkevich, the Soviet commander of Russian aircraft in Spain for the first part of the war, had suggested building *Chatos* in Spain. However, production was slow and only forty-five machines were built in 1937 and only four in February 1938 because the enemy continually harassed the factories.[9] Although a total of 237 was completed by the numerous factories of the *Servicio de Aviación y Fabricación* before the end of the war, they may not all have flown, because the number of engines sent from the USSR altogether was 269 and these were to replace worn-out existing ones.[10]

In an effort to redress the growing imbalance in both quality and quantity between the Republican and Nationalist air arms, in April 1938 the USSR sent the Spanish Republic thirty-one I-16 Type 10s powered by more powerful engines, followed by a further ninety in June and July, but the Soviet pilots who flight-tested the aircraft reported several alarming faults. Defects in the construction of the engines, as well as in the armament, caused a general reluctance among pilots to fly combat missions in the new fighters. Nevertheless, the new engine gave this aircraft a maximum speed of 275 mph at 9,000 feet and the power to climb to 15,000 feet in seven minutes. In addition, it was armed with two additional machine guns, firing between the propeller blades. For a time, the Republic had seven more or less complete squadrons of *Moscas*. In April 1938, however, the Germans reacted and dispatched five Messerschmitt Bf-109C-1s which had a greater maximum speed, though a slightly slower rate of climb, and two additional machine guns. That summer, five Bf 109Ds, a later version of an eventual 35, were delivered. Even later versions, the Messerschmitt 109E-1 and the E-3, forty-five of which would arrive in December 1938 and early in 1939, made a total of 131 of these superb fighters delivered to Franco's air force during the Spanish Civil War.[11] Together with the Italian CR-32 *Chirri*, with its robustness and remarkable powers of manoeuvre, added to the professionalism of the pilots, the Nationalists enjoyed fighter superiority most of the time.

On 30 October 1937, the Condor Legion had received its next commander. General Sperrle was replaced by General Helmuth Volkmann (alias 'Veith' in Spain). The command staff were housed in their *Wohnzug*, a twelve-carriage train protected by a *flak* battery, which could be moved easily from place to place, so long as the railway reached to wherever the Condor Legion headquarters was required. Now it was at Almazán, north of the Sierra de Guadarrama so that it was equidistant from Madrid and Zaragoza. However, Franco, triumphant in the north of Spain but having tried many times without success to take the capital, was by now determined to push east into Aragon and perhaps to reach the Mediterranean.

Conditions were difficult. The Legion badly needed a rest. Supplies of spare parts were lacking and it seemed that Berlin was losing interest in the Spanish war. One of the German pilots, the later Colonel Pitcairn, told the American military historian Raymond Proctor:

> We had the feeling that we had been sent to Spain and then deserted [. . .] We had to bale ourselves out of a generally forgotten war.[12]

Nevertheless, new aircraft continued to arrive from Germany. The fighter group was reformed into two squadrons of He-51s, still very useful for close-support missions though not for tackling Soviet fighters, and two of Messerschmitt Bf-109s. All four bomber squadrons were equipped with the up-to-date Heinkel-111s. Older aircraft were passed on to the Spanish Nationalists. A seaplane squadron was based on the island of Majorca, invaluable for attacking shipping on its way to Republican-held ports along the Mediterranean coast. The anti-aircraft or *Flak* group had four batteries of the famous 88mm cannon, two of the 20 mm and one of 37 mm.[13]

Teruel

The next major Republican operation was the capture of the Nationalist salient of Teruel in Aragon. The offensive began on 15 December 1937. This was the worst possible time, given that Teruel has some of the harshest winter weather in Spain. It may be that the government wanted to forestall a rumoured Franco offensive against Guadalajara and thence to Madrid, or that the time seemed ripe for a Republican victory which might force Franco to agree to a peace or persuade his German and Italian supporters to abandon him.

The initial Republican advance was successful, and by evening on 15 December 1937, Teruel was surrounded by three Republican army corps, but as nearly always in the Spanish war, it took days to penetrate the town itself, for the Republican units were composed to a large extent of poorly prepared, ill-equipped and amateurly led new conscripts. For once, however, Franco's reaction was slow. His counter-attack began on 19 December, protected by the Condor Legion. However, within a day or two, the weather closed in. Temperatures fell to 18 degrees below freezing and no motors could function. Both Franco's forces besieging the Republicans and the latter, who were in turn besieging the last defenders of Teruel, were cut off from their supply lines by heavy blizzards. Hundreds of lorries were stuck in snowdrifts and men were losing limbs to frostbite. On 8 January 1938, the defending Nationalists, holed up in the thick-walled buildings of the old town, surrendered. Now the Republicans inside the city became the besieged. Between 4 and 7 February, Franco's forces launched an attack north of Teruel towards the river Alfambra. The Republican forces, weakened by fatigue, lack of supplies and occasional mutiny, and faced by irresistible Nationalist superiority of artillery and air firepower, collapsed and retreated across the river. Teruel was encircled and finally retaken by the Nationalists on 20 February.

To prepare for their assault on Teruel, the Republican air force had transferred four squadrons of I-16 fighters to aerodromes in the Mediterranean provinces of Valencia and Castellón. These were led by the Spanish pilots Eduardo Claudín, Chindasvinto González, Juan Comas and Manuel Zarauza, who had been trained during the war in Spain or in the Soviet Union. There were also four squadrons of I-16s under Russian command and three squadrons of *Katiuska* bombers.

General Volkmann, commander of the Condor Legion, together with the Italian military chiefs, was critical of Franco for concerning himself with relieving Teruel when, in their view, the planned attack towards Guadalajara and on to Madrid would have succeeded in taking the capital and ending the war. Nevertheless, the command train – the *Wohnzug* – got up steam and moved to Calamocha, 70 kilometres north of Teruel. He-46 reconnaissance and low-level bombers flew in while other air units swiftly moved to airfields whence they would be able to operate over Teruel. Eckehart Priebe, an officer with the new 4th J/88 fighter squadron, recalled:

> At León we [. . .] assembled our Heinkel-51s and off we went to the Guadalajara [he probably meant 'Guadarrama'.M.A] Mountains for another of those 'final assaults on Madrid' which never took place. Instead, on Christmas Eve 1937, we had to hurry to a field called Calamocha, south of Zaragoza, to help the beleaguered garrison at the city of Teruel which the Reds had surrounded in a surprise offensive. For more than two months we were engaged in a most bitter battle in ice-cold Aragon. We flew three or four ground-support missions a day at very low level, strafing trenches or dropping our six 10 kg bombs on gun positions or military transport. Aerial combat was left to the Bf-109s, as the Heinkels were no match for the Soviet-made Ratas [I-16 *Moscas*].

Captain Gotthardt Handrick, in command of the fighters, recalled:

> None of us had believed it to be possible that in Spain it could become really cold. The scorching heat of the summer had, in fact, helped to give us the notion that we would also experience a warm and pleasant winter. Shockingly, when I made my first reconnaissance flight to the Teruel Front on Christmas Eve, the thermometer showed minus 18°C. It was especially difficult for our ground crews at Calamocha. Night after night and hour after hour, a special group of ground staff had to rev the aircraft up in order to keep them warm.

Exact figures from military sources for 15 December, when the battle of Teruel began, indicate that there were 460 Nationalist aircraft, of which 210 were of the *Aviazione Legionaria*, 100 belonged to the Condor Legion and 150 mostly older machines to the Spanish air force, though eight of the latter's best squadrons were equipped with Italian aircraft.[14] Flying conditions were extremely difficult, but Franco expected the Italians and the Germans to support his infantry corps, so Italians and Germans flew whenever there was a break in the weather. Then the heavy snowstorms and arctic temperatures which made road transport impossible also paralyzed air operations. Engines refused to start; propellers could not be moved. When in the end, by using blowtorches on the cylinders and chipping ice off the wings, the Condor Legion managed to put its machines into the air, they were forced down by fierce winds and driving snow. On 17 December 1937, twenty-three Heinkel-111 bombers took off and were scattered by a blizzard. One of them landed intact in enemy territory. When the Russians insisted that it be sent to the Soviet Union to be inspected by experts, Indalecio Prieto, Minister of National Defencefor Navy and Air, refused and, hoping for the French frontier to be opened to allow arms purchases to be brought into Spain, invited a team of French officers and aeronautical engineers to come and fly it. Only then did Prieto allow the Heinkel, together with a Messerschmitt Bf-109 which had also been captured intact, to be shipped to the USSR.[15] The German army recovered the fighter during the Second World War.[16]

It was at this moment that Colonel Wolfram von Richthofen was recalled to Germany and replaced as chief of staff by Major Hermann Plocher, who had been the original organizer of the Condor Legion. The personalities of *Generalmajor* Hellmuth Volkmann and *Oberstleutnant* Dr.-Ing. Wolfram Freiherr von Richthofen had clashed following simmering tensions. On 11 January, von Richthofen noted, 'I request my release [...] immediately, some leave here and then a return trip home.' Two days later, he wrote to his wife, 'Volkmann and I must part company at the soonest possible date.'[17] By 30 January 1938, he had gone.

As the weather improved, the Nationalists advanced under cover of the Condor Legion fighters and Heinkel-111 bombers. Six Italian Fiat BR-20 fast bombers flew missions high over Teruel at their high cruising speed which allowed them to operate without fighter opposition. Fiat *Chirris* escorted He-45 and He-46 low-level bombers, attacked by *Chatos* protected by *Moscas*. German aircraft, piloted either by the Condor Legion or Spaniards, dropped 120 tons of bombs over the Teruel battlefield on 6 January 1938, even more than when they had bombed the 'Iron Belt' around Bilbao in June 1937.

When Soviet *Katiuskas* were ordered to bomb the Nationalist-held city of Salamanca, the Minister of National Defence, Indalecio Prieto, remarked that they would have been better used at Teruel.[18] Some certainly were, for when, identifying a weak stretch of the Republican line north of Teruel, Franco had launched a counter-offensive in this area – a flanking attack with a spectacular cavalry charge that helped turn the tide and trap the Republicans in Teruel – three of the fast *Katiuskas* fell to the guns of the second German fighter squadron (2.J/88) within six minutes. Captain Gotthardt Handrick recalled, with an element of over-claiming:

> None of us would have had good memories of Calamocha had it not been for 7 February 1938. This was the date of a quite special triumph for my *Gruppe*, for we were successful in shooting down no fewer than 12 enemy aircraft – ten Martin bombers [SB *Katiuskas*] and two *Ratas* [I-15 *Moscas*]within five minutes.

It was Lieutenant Wilhelm Balthasar, flying a Messerschmitt Bf-109, who shot down these fast Russian bombers and, on his way back to base, downed a *Mosca* fighter. This last would be his seventh and final victory in Spain. Perhaps this Condor Legion ace thought that the red scarf that he had taken from the neck of a dead Republican and worn ever since, was his talisman.[19] It would not save him from death in action in 1940.[20]

On 2 January 1938, while escorting German and Italian bombers over the Teruel battlefield, a young Spanish second lieutenant, José María Careaga, flying a *Chirri*, was shot down by intense small arms and anti-aircraft fire. He was taken to the headquarters of the Republican 70th division, commanded by Hilamón Toral, a militia officer who had run a boxing academy in Bilbao and who recognized Careaga at once as an ex-pupil and entertained him. Careaga spent the next year as a prisoner in Barcelona until the Catalan capital was taken by Franco on 26 January 1939.[21]

It was during the battle of Teruel that the famous or perhaps notorious, Junkers-87 dive-bomber or *Stuka*, with its screaming siren, made its first appearance. One of these planes, which would harass columns of refugees and retreating soldiers in France in 1940, was sent to Spain for evaluation earlier in the civil war but later sent back to Germany. Three more advanced *Stukas* arrived at the Condor Legion's headquarters at Vitoria on 15 January 1938. They were transferred to Calamocha on 7 February and played some part in the final stages of the battle for Teruel and in later combat in Spain, though in conditions of such absolute secrecy that not even General Franco himself, while officially visiting the Condor Legion, was allowed into the hangar in which the *Stukas* were housed. Later in 1938 the three *Stukas* were returned to Germany and replaced by three more of an advanced version. This was the Ju-87b, the classic *Stuka* of the beginning of the Second World War. These dive-bombers took part in the final battles in the Spanish Civil War and were returned at once to Germany, without, interestingly, making their appearance at the great victory air display at Barajas on 12 May 1939.[22]

Franco cuts the Republic in two

In the spring of 1938, an approximate comparison of aircraft available to both sides, according to a competent, leading historian of Francoist sympathies, was as follows:[23]

The Republic had four squadrons of I-15 *Chatos* and six of 1-16 *Moscas*, plus a squadron defending the major port and naval base of Cartagena, giving a total of about 120 fighter planes. The Nationalists had thirteen squadrons of Fiat CR-32 *Chirris*, and two of German Bf-109s, with a total of 140–150 fighters.

The Republic had fifty SB-2 *Katiuska* bombers and an unknown number of other bombers such as Marcel Bloch, DC-2 and Fokker-VII, while the Nationalists had twelve Italian S-79s, twenty-one S-81s, twelve He-111s and five Dornier-17s, added to fifty-six Ju-52s, a total of 106. Here the historian fails to mention the Grumman and odd Douglas and Fokker aircraft that one side and the other had had since the beginning of the war.[24]

How available spare parts were and how many aircraft were in a condition to fly missions at a given moment, as well as the level of experience and training of the pilots, are questions which bald numbers of aircraft do not answer. Whatever the valour of the Republican pilots, by now they were mostly the product of rapid training courses in Spain or the Soviet Union The second cohort of Spaniards trained in the USSR returned at Christmas 1937 and, after leave, were sent for additional training at El Carmolí in the province of Murcia before being sent to the 1st and 3rd squadrons of I-16 *Moscas*, under the command of Eduardo Claudín and Manuel Zarauza, and even their squadron leaders and wing commanders had had relatively little flying and no leadership experience before the war. The new pilots were given extra training but this was never long enough, given the losses. As soon as ready they were sent to the front, leaving the coastal area, especially ports such as Barcelona and Tarragona, at risk from Italian bombers flying from Majorca. The Nationalists had no need to defend their own ports, which were far away in north-western Spain, at a distance where Republican bombers could not be escorted by fighters. The new pilots, in Lacalle's view, were insufficiently trained, had little combat experience and little self-confidence. In particular, Lacalle was disturbed by their underestimation of distance, which led them to open fire and exhaust their ammunition while still too far from their target. He was also worried by their lack of the essential spirit of aggression and initiative. They lacked complete control of their aircraft and had not been trained to push them to their limits in climbing and diving. Lacalle's comments seem to carry a nuance of criticism of unimaginative Soviet training. Without sufficient machines with dual control, it was impossible to correct these deficiencies once the pilots were back in Spain. Similarly, there was just not enough time to train bomber pilots to use the *Katiuska* to its best advantage, even though it was practically untouchable by a fighter unless the latter was higher and came down on the *Katiuska* by surprise, as indeed skilled Nationalist pilots sometimes did.[25] The more recently arrived Soviet military pilots, for their part, had accumulated far fewer flying hours than the Germans and Italians. They lacked training in take-offs and landings at emergency airfields. Some were incompetent at navigation. All these deficiencies emerge from statistics of the first two years of the civil war, during which 163 Soviet aircraft were lost in combat in the air, but 147 were lost in forced landing and accidents. Francoist aircraft were rarely lost in this way.[26] Furthermore, the German bombers and the Messerschmitt Bf-109 fighters were superior to whatever the Republic could put into the air, and the Fiat CR-32 *Chirri*, in the hands of pilots such as García Morato, was a robust match for the *Chato*. As for the Soviet *Katiuska* bomber, it had many defects, particularly its unprotected fuel tanks, which made it inferior to the latest Italian bomber, the Savoia-Marchetti S-79, and the German Heinkel-111. Moreover, the technical superiority of German anti-aircraft fire has to be considered. The Republican artillery often found that its shells were dud and it may well be that the absence of professional direction of artillery fire in the Republican anti-aircraft batteries played an important part.[27]

However, now that the Nationalists had imposed a virtual blockade of the sea route from the USSR, in the autumn of 1937 only 50 aircraft arrived for the Republic in contrast to the 272 received by the Nationalists. Although in the first quarter of 1938 the Republic would receive eighty-nine planes and the Nationalists only thirty-six, the latter were now able to dominate the skies.[28] Furthermore, there was no shortage of highly trained professional Italian and German military pilots anxious to obtain experience and the chance of promotion in Spain, whereas the Soviet Union, eager to prove to the Democracies that it was not interfering in Spain, was steadily withdrawing its aviators. On the eve of the battle of Teruel, in November 1937, the *Aviazione Legionaria* had a force of 2,008 men, including 327 pilots. In comparison, German numbers of men in Spain peaked at about five thousand, probably because there were more ground crews and men to man the anti-tank and other auxiliary units, but the number of planes was rarely more than one hundred at any given moment. So, if the number of aircraft is the major criterion, Fascist Italy contributed significantly more than Nazi Germany to Franco's victory in the air, not to speak of the Italian army corps, the CTV, which fought in Spain.

Majorca

Not least of the Italian contribution was the role of their aircraft on the island of Majorca, largest of the Balearics. From the beginning of the war, Italy had realized that Majorca was an ace in the insurgents' hands. It would in time be an important base for Franco's navy and the Italian submarines which threatened the Republican navy, as well as a base for seaplanes to harass shipping making for Republican ports along the Mediterranean coast. If a victorious Franco could be prevailed on to cooperate, an Italian presence on Majorca after the Spanish war would offer a real threat to British and French naval activities in the Western Mediterranean. Mussolini's victory in taking Santander in August 1937 encouraged him to take an aggressive stance by reinforcing his air force on Majorca with the faster and better armed Savoia-Marchetti S-79s. Twelve of these, probably the best bombers to fly in Spain, landed on the island on 27 September 1937. One of them was piloted by Bruno Mussolini, the dictator's son. A few months later, Rómulo Negrín, son of the Spanish Prime Minister, would begin to fly fighter missions on the other side. These bombers, vaingloriously titled 'Falcons of the Balearics' (*Falchi delle Baleari*), bombed Alicante on 30 September 1937, Valencia on 3 October and Sagunto on 6 October. Even the *Chatos* that were transferred from the front to tackle the S-79s found that the Italian bombers came in from Majorca very early in the morning, with the sun behind them so that they could not easily be spotted until they were almost over their target. By the time the Republican fighters could climb high enough, the Italian bombers were far away flying back to their base on Majorca. Republican anti-aircraft defences managed to down only two S-79s between October 1937 and November 1938.[29]

During the battle of Teruel, about thirty Italian aircraft from their bases on Majorca bombed strategic targets along the Mediterranean coast of Spain, a method of air warfare which the Italian theorists would have preferred to the ground-support tactics favoured by the Germans. In time they would get their chance to launch heavy strategic bombing raids on Valencia and Barcelona. Reflecting his feeling that Franco urgently needed Italian support, on 2 February 1938, General Berti, the new commander of the CTV, handed Franco a letter from Mussolini threatening to withdraw Italian support unless he concentrated all his forces to take Madrid. Franco replied in a conciliatory tone, underlining the importance for him of Italian aircraft, which had indeed played the major part in the successful recapture of Teruel.[30]

Franco drives east to the Mediterranean: The Republic split in two

With the north of Spain in Franco's hands, and with its mines, industrial plant and tens of thousands of prisoners who could be 'recycled' at least as auxiliary troops, the Spanish insurgent general was ready to make a dash eastwards from Aragon and perhaps take Valencia, Spain's third city, which had become its temporary capital since the government had hurriedly fled Madrid in November 1936 when it seemed likely to fall. While on the one hand it seemed possible that the Democracies would at last stand firm and dissuade Germany and Italy from aiding Franco, on the other the resignation on 20 February 1938 of Anthony Eden, the British Foreign Secretary, seemed to make it clear that, against his advice, the British government would accept the presence in Spain of Italy's massive contingent of men and aircraft. Franco, confident that Hitler and Mussolini would replace his losses, could now accumulate irresistible masses of aircraft and artillery, against a front which he knew was likely to collapse.

The breakthrough began on 9 March 1938, only two weeks after the end of the battle of Teruel. Perhaps Franco feared that the planned *Anschluss*, the German annexation of Austria, would draw the Condor Legion away from Spain. Indeed, on 12 March, Germany annexed Austria. However, there was no British reaction, while the French government resigned. Camille Chautemps was replaced by Léon Blum, who had been dissuaded in the summer of 1936 from sending any significant amount of war material to Republican Spain. He now promised to relax controls on the frontier with Catalonia but at the same time could do little because of British opposition and the hesitations of his own ministers and the French military leaders. On 15 March, Britain signed an agreement with Italy, which, at least tacitly, led Mussolini to assume correctly that Britain would tolerate the presence of Italian troops and aircraft in Spain until Franco's ultimate triumph. Franco thus had no reason to fear that his allies would leave him in the lurch.[31]

Beginning with an artillery barrage at 5.30 in the morning of 9 March, Franco's four army corps drove through Aragón, using a *blitzkrieg* technique with tanks advancing under the protective cover of fast Heinkel-111 bombers which attacked strong points, reserves, depots and columns of troops on foot or in trucks. Low-flying squadrons of He-51s made repeated strafes in front of advancing Nationalist troops. Without losing a single bomber, the advancing Nationalist troops destroyed six Republican divisions, which sometimes panicked and could not be controlled by their poorly trained leaders.

On day two of the attack, the Condor Legion directed its fire on to enemy airfields, each aircraft flying between three and six sorties.[32] In a week, Franco's divisions, covered by the Italian fighter wings *La Cucaracha*, *As de Bastos* and *Gamba de Ferro*, and another three flown by Spanish Nationalist pilots who escorted the Ju-52 bombers, advanced 70 miles from their jump-off points, occupying 7,000 square kilometres of territory, capturing huge numbers of prisoners and abandoned equipment, forcing the Republican fighter squadrons to move hastily to airfields in the rear and obliging the Republic's army staff to reorganize its brigades and divisions feverishly as it strove to collect dispersed groups of panic-stricken men. Republican defence of their airfields was evidently insufficient or the aircraft were not properly camouflaged because, on 9 March

Nationalist bombers destroyed two *Chatos* on the ground at Caspe, and on 10 March four *Chatos* on the ground at Lérida. On 11 March, German fighters destroyed another five parked *Chatos*. On the same day, two more *Chatos*, flown by Russians, lost their way or perhaps fled into France.[33] On 24 March, seven more Chatos were lost in combat.[34] The Republic lost sixty aircraft through enemy action during the offensive in Aragon – mainly I-15s and I-16s – half of which were destroyed on the ground or abandoned during the retreat. Several others were lost through accident.

While the Nationalist *Chirris* flew from an aerodrome near the recaptured town of Belchite, the Condor Legion flew its He-51s and Bf-109s to Sariñena, the aerodrome where the 'Red Wings' squadron from Barcelona had been based during the Republican militia's march into Aragon in August 1936 and had now been evacuated by the *Mosca* squadrons which had flown from there. From Sariñena the Condor Legion's staff, its fighters, its short-range He-45s and the Ju-52s, which flew the four weekly courier trips to Germany, moved to La Cenia, north of Vinaroz and just inside the province of Tarragona which would be their permanent base for much time. This was only 10 miles from the coast. The Germans remained at La Cenia for several weeks. Here, at last, they could enjoy warm weather tempered by sea breezes. Swimming, films flown in from Germany, concerts by the Legion's band and fraternization with local people made the weeks in La Cenia very pleasant.[35] On 2 June 1938, nine *Katiuska* bombers set off to raid this German base. It was heavily defended with 88mm *flak* artillery, which immediately scattered the bombers and left them prey to the Messerschmitt Bf-109s. Five *Katiuskas* were lost, two by impacts from the 88mm guns and three downed by the Messerschmitts. The surviving Republican pilots were transferred to gaol in Salamanca where they joined seventeen others, as well as three Russians and the American Harold Dahl. Here they would be held pending possible exchange for Nationalist prisoners.[36]

During the weeks of Franco's race to the sea in spring 1938, the *Aviazione Legionaria* put an average of 170 aircraft into the air daily.[37] One of the squadrons of Italian fighters, almost completely composed of Spanish professional military pilots, was led by the ace Captain Angel Salas. On 8 April, now promoted to major, having escorted two bombers to the front, his aircraft was hit and lost all its oil. The ground was rough and Salas had already released the lock on his parachute when he saw a piece of level ground, relocked his parachute and skilfully landed his fighter, with damage only to the lower part of the left wing and the landing gear. Salas flew the same Fiat for most of the war, starting with his service with the Italian captain Dequal's squadron in 1936. He recovered it when the squadron became completely Spanish as part of 2-G-3 (2nd wing of Fiat CR-32 fighters) and flew it throughout 1937, and from January to April 1938. After repair, he flew it once more from July to November. After an accident, which kept the fighter under repair, Salas flew it from January 1939 until the end of the war and for long afterwards, until it was destroyed by a major accident.[38]

Condor Legion bombers carried out 'Operation Neptune' on 16–18 April, with forty He-111 bombers. They took off from Sanjurjo aerodrome near Zaragoza. In order to avoid Republican territory, they flew west to either Avila or Salamanca to refuel before keeping to Nationalist air space all the way south along the Portuguese frontier into Andalusia as they made for the Republic's naval base of Cartagena and the port of Almería on the south-east coast of Spain, where Soviet supplies were being landed. Low cloud and an error in navigation caused one aircraft to crash into a hillside in Extremadura, killing the five crewmen. Another made an emergency landing because of problems with the fuel links. Two more returned to base because of mechanical problems and other causes prevented two more from completing their journey. That evening, thirty-four He-111s out of the forty assembled at Armilla aerodrome near Granada. From here they crossed the Sierra Nevada, whose heights are snow-covered even in summer, and came in over Cartagena. They hit twenty-four ships at anchor, the railway and the Campsa oil refinery. Returning to Granada to load bombs and refuel, on their second raid they sank a torpedo boat and also did a lot of damage in Almería. Although Cartagena, the Republic's principal naval base, was well-defended, only one bomber was hit by anti-aircraft fire. Its crew attempted to fly along the coast, but it was forced to ditch at Motril, close to the junction of the two zones of Spain. Other planes were forced to land at Málaga, and on 18 April, only twenty-seven Heinkel-111s of the forty which had taken off returned to their base near Zaragoza. The Condor Legion command was depressed by such a poor result for such serious losses.

These He-111Bs were transferred to the Nationalists in June 1938 and replaced by He-111E1s, whose larger engines and greater streamlining allowed them to reach a maximum speed of 260 mph at 12,000 feet. Altogether, between seventy-five and one hundred He-111s were delivered to the Nationalist air force. They were heavily used and some forty were lost in combat and accidents. It was Spanish-built Heinkel-111s, repainted in the colours of the Luftwaffe, which would be used for the 1969 film *The Battle of Britain*.[39]

On 20 April 1938, Franco's armies, sometimes advancing as much as 28 miles in a day, reached a line stretching from the French frontier south through the Catalan towns of Sort, Tremp and Balaguer on the Noguera-Pallaresa river, which flows into the Segre, and along the latter until its junction with the Ebro between Tortosa and the sea. All those towns where collectives had been established by the Council of Aragon in the distant, heady days of anarchist revolution fell in a few days. On 3 April, Lérida was the first provincial capital of Catalonia to fall. Franco's government immediately abolished the Catalan Statute of Autonomy, while the inhabitants were ordered to speak Castilian Spanish and the town would henceforth be known in all official documents as Lérida rather than the Catalan Lleida.

On 6 April 1938, following public demonstrations against suggestions that an armistice should be sought, the Republican government was reorganized. Indalecio Prieto, fiercely attacked for what many saw as his defeatism but which he considered a realistic stance, was removed from the Ministry of National Defence. The Spanish Communist Party, now seen, especially because of Russian aid, as the maximum defender of the Republic against Franco, clashed with Prieto also because the latter wanted a strong independent air force which, whatever the help that had to be accepted from Moscow, would in no way be subordinate to Soviet needs or the orders of the Soviet advisers in Spain. If the Republic survived the war, it was feared, probably quite unrealistically, that its air force would be ideologically communist because so many of the cohorts of trainees at Kirovabad, indoctrinated in the USSR, were to become commanders of wings, and squadrons.

The tension caused a split between Prieto and the commander of the air force of the Republic, Ignacio Hidalgo de Cisneros, who had been personal friends. The Ministry of National Defence was now taken over by the Prime Minister, Juan Negrín, while Colonel Antonio Camacho was replaced by Colonel Carlos Núñez Maza as Undersecretary for Air. To raise public morale, one of the latter's first acts was to order a spectacular demonstration on 9 April by sixty planes over Barcelona, to which the Republican government had moved from Valencia.

On 15 April, 1938, Good Friday, Nationalist troops had reached the Mediterranean at Vinaroz, just south of where the Ebro reaches the sea. The disaster was complete. The Republic had been split in two.

The obvious movement for Franco to make now was to drive northwards towards Barcelona and close the French frontier to any possibility of armament coming by the land route, thus encircling what would be left of Republican Spain, but Franco feared a fierce reaction from the Blum government if his forces, aided by Germany and Italy, reached the Pyrenees. He turned south towards Valencia. However, the approaches to the city were strongly defended and by mid-June 1938 the lines had stabilized some 68 miles north of it.

Meanwhile, Republican ports along the Mediterranean coast were being heavily bombed by Italian bombers taking off from Majorca. The spring of 1938 saw a new and more terrible aspect of the Spanish Civil War. Italian theorists were much influenced by the ideas of Giulio Douhet who thought that a war could be won merely by bombing the civilian populations of large cities so intensively that they would demand an end to hostilities. Early bombing in Spain, as has been seen, was ineffective and tended to produce reprisals against prisoners. Bombing Madrid was also considered counterproductive because it might easily harm Nationalist sympathizers as well as the enemy. However, the presence in Spain of fast, modern bombers led to their use in the strategic bombing of installations such as power stations, railway yards and port installations, for which the Heinkel 111s and the Savoia-Marchetti S-71s were fitted.

To bomb from great height the Republic had only the SB-2 *Katiuska*, a vulnerable aircraft which required fighter escort. For over a year, there was little Republican bombing of towns. After the Italians began bombing Barcelona, Reus and Tarragona, Indalecio Prieto, Minister of National Defence until early April, 1938, issued an official note:

> Facing the bomber [...] there is only one recourse: the bomber, used in the same way as our enemies use it, but in greater numbers if possible. In other words 'Terror against Terror'.[40]

Yet this was not reflected in the Republic's bombing of enemy towns. Córdoba, for instance, a major railway junction in southern Spain, was bombed frequently but by small numbers of aircraft. The casualties, though tragic, were never very great compared with those suffered in Madrid, Valencia and Barcelona, the three biggest cities still in the hands of the Republican government.

The destruction of Guernica on 26 April 1937 can be seen as an experiment in terror-bombing, but it was over a small area and had some tactical importance in that it was aimed at cutting off the retreat of Basque troops. Barcelona, however, Spain's second city, was bombed a total of 113 times by Italian bombers, sixty times by the Condor Legion and once only by the Spanish nationalists. The raids caused 2,500 deaths.[41]

The inhabitants of Barcelona were used to raids on the port. However, on 16 March 1938, Mussolini summoned General Valle, Italy's Undersecretary for Air, and ordered him to carry out a massive bombing raid on Barcelona. 'Begin a violent action on Barcelona from tonight' (*Iniziare da stanotte azione violenta su Barcelona*).[42] Consequently, from eight minutes past ten on the night of 16 March 1938 until three o'clock in the afternoon of 18 March, extensive residential parts of the city were bombed by thirteen waves of Italian Savoia-Marchettis at three-hourly intervals. Barcelona had little anti-aircraft artillery and no fighter cover. Andrés García Lacalle, now responsible for coastal defence, asked for the urgent presence of *Mosca* fighters but these did not reach the Catalan capital until the morning of 17 March, perhaps because they were urgently needed at the front. The Italians, emboldened, carried out bombing attacks from under 3,000 feet.[43]

Mass international protests followed. Eberhardt von Stöhrer, German ambassador to Franco, sent a long cable to Berlin, reporting that the results of the bombing were 'nothing less than terrible'. The ambassador stated that all parts of the city had been affected and there had been no attempt to hit military objectives in particular. Hundreds of houses and entire districts had been destroyed. Women queuing for food and people taking shelter in the underground railway stations had been killed. Von Stöhrer seemed rather happy to point out that as the wounded were carried off to hospitals they defiantly gave the clenched fist salute. The German ambassador feared that after the war the memory of the bombing would stir up hatred for the Italians and the Germans. He was happy to say that he had just heard that Franco had forbidden any more such bombings of civilian targets.[44]

In reality, the horror of the Italian bombing of Barcelona and other cities in Republican Spain hides the fact that, in comparison with the large casualty list at Guernica, progress had been made in air raid precautions. These were closely studied in Britain, particularly following a book written by Professor J. B. S. Haldane, of University College London, who had investigated the bombings in Spain.[45] The book was reprinted twice in its first month.

Valencia was first bombed on Saturday, 15 May 1937, when eight S-79s flew the 164 miles from Palma to the temporary capital of Republican Spain. In the early evening, they bombed along a hundred-metre wide corridor from the low height of 1,250–1,850 feet. The bombers glided in unnoticed by the sound detectors, dropping high explosive and some incendiaries which proved defective, leaving thirty-two dead and seventy-two wounded, of whom many died. A few days later Valencia was bombed once more, while on 26 January 1938 six bombers dropped forty-two bombs and left over two hundred victims. After Barcelona had been the target in March 1938, the Italian air force in the Balearic Islands tried to enforce an air blockade of the ports of Almería, Barcelona, Rosas, Tarragona, Sagunto and Valencia, in the hopes of strangling the Republic's imports of raw materials and foodstuffs, as well as its essential exports.

Radar to give advance warning of the approach of bombers, and ground control systems to guide fighters towards approaching raiders, would have to wait until the Second World War. In Spain there were only ground observers, sound detectors and telephones so that by the time the fighters took off and reached the appropriate height, the bombers had dropped their bombs and were racing home at what was by 1938 often a speed that fighters could not match. This would not last, however. The increasing speed of fighters such as the Bf-109 and their powerful armament (with cannon rather than machine guns) would require bombers to be escorted. By the Second World War, while theorists had predicted two million casualties if London were bombed, in Great Britain as a whole the entire number of air raid casualties had been three hundred thousand, half of which were slight injuries, so the alarmist statement of the British Prime Minister Stanley Baldwin that 'the bomber would always get through' was not correct. Nor were Douhet's forecasts accurate. Heavy bombing did not destroy civilian morale, either in Spain or in Britain, especially if adequate shelter was provided. And even the continuous day and night bombing of German cities by the RAF and the US Army Air Force towards the end of the Second World War would not substantially impede war production.

At this point, June 1938, with the Republican armies in chaos, Franco could advance again on Madrid. He would be surprised, however, by the extraordinarily swift recovery of the wrecked Republican armies and their capacity to launch a new attack from Catalonia over the river Ebro.

Chapter 9
The battle of the Ebro: Republican armies retreat through Catalonia

The Republican armies, under the efficient and energetic direction of Juan Negrín, Prime Minister and Minister of National Defence; Colonel Antonio Cordón, Undersecretary for the Army; and Colonel Vicente Rojo, chief of staff, recovered remarkably swiftly from the disaster of March and April 1938. The briefly reopened French frontier allowed for a mass of equipment to enter Republican Spain. The Army of the East was reformed and a new army, called 'of the Ebro', was prepared to launch an attack across the Ebro river.

The Army of the Ebro was the elite of the entire Republican army. It was communist-led, commanded by the only Spanish militia officer to become a general, Juan Guilloto León, always known as 'Modesto', and was composed of three army corps under the experienced and determined militia officers Enrique Líster, Manuel Tagüeña and Etelvino Vega. Well-disciplined and trained, the Army of the Ebro was a match for Franco's best troops. Nevertheless, it still suffered from its lack of career officers at lower levels, a deficiency not compensated by a rigid command structure.

At a quarter past midnight, on the night of 25–26 July 1938, the first units of the Army of the Ebro crossed the river on a great bend between the towns of Mequinenza and Amposta, in some of the hundreds of flat-bottomed boats which had been constructed and brought down to the banks of the river, apparently without the knowledge of the enemy. Once across, footbridges and pontoon bridges were rapidly built to allow troops to cross in large numbers. The next thing would be the building of an iron bridge capable of carrying tanks and trucks, before the enemy opened the dams higher up on the River Segre.

Success was immediate. Nationalist troops retreated and entire units were captured. Franco, however, reacted at once, using his efficient logistics to move divisions from other fronts. The 102nd Division, for instance, left Mérida in south-west Spain on 27 July and by dawn on 30 July it was in line digging trenches.[1] The Nationalist fighter pilot José Larios, whose fighter group moved from Extremadura to the Ebro, recalls that the Nationalists sent skeleton ground crews by road overnight to meet the fighters.

> To move a fighter group at only a few hours' notice is a complex matter which calls for a high degree of training and efficiency. It involved mobilising a large number of ground personnel and masses of spares, baggage, tents, fuel, trucks, buses [...] loading them up at the double and putting them on the road within a matter of three to four hours.[2]

Franco ordered concentrated reaction by his air force, which by the following day was in action, using the 'chain' or *cadena* technique of near-continuous dive-bombing of the footbridges to prevent the passage of men, vehicles and artillery. Italian bombers were also used to destroy the bridges. For example, on 9 August, S-79s bombed the bridge at Flix twice, once at noon and once at dusk, with nine and five aircraft, dropping 120 bombs. They repeated the bombing on the 11 August, when six aircraft dropped thirty-five bombs.[3] Italian pilots tended to bomb from great heights, so it may be that few of the bombs found a target. Furthermore, the Nationalists opened dams higher up the river Segre and many of the footbridges were swept away. So, while advanced Republican units had quickly reached the outskirts of Gandesa, some 7 miles from their jump-off points at Flix and Ascó on the north bank of the Ebro, they found themselves unable to move further.

One week after its launch, the Republican attack over the Ebro had failed. There was not going to be a breakthrough. The battle would become a slogging match over the next three months, as the Republican brigades and divisions would struggle to defend the large pocket they had created and Franco would wear them down with the advantage of his larger numbers, his heavy artillery and his much larger air force, particularly his low-level bombers. When Léon Blum's government fell in April 1938, the French frontier was closed. Thus the Republic, with its Mediterranean ports blockaded and under attack from German and Italian seaplanes based on Majorca, would not be able to import war material to the same extent as Franco could call for armament from Germany and Italy. Concentrated Nationalist artillery and air bombardment against troops unable to dig deep shelters in the rocky terrain of the mountain ranges, to which they desperately clung, would tell in the end.

On 30 September 1938, as the battle of the Ebro was at its height, France and Britain signed the Munich agreement, appeasing Hitler's demands on

Czechoslovakia. While on the one hand it seemed that there was no more hope that the Democracies would come to the aid of the Spanish Republic by allowing it to acquire arms freely, on the other hand Juan Negrín, the Republic's war leader, was convinced that sooner or later tension would rise again and the Spanish war would be subsumed in a major conflict between the Democracies and Germany and Italy. For Negrín, this meant that resistance to Franco was essential. Both Spanish sides made moves to court the favour of the Democracies, Franco by withdrawing ten thousand Italian troops, but of course retaining the much more valuable Italian aircraft and pilots, and the Republic by removing the last twelve thousand men of the International Brigades. Nevertheless, relentless Nationalist pressure forced the Republican armies to cross back to the other side of the Ebro in mid-November 1938. Though badly mauled, they crossed in good order, which enabled resistance to continue in Catalonia for a further two and a half months.

The aircraft available to both sides at the beginning of the battle of the Ebro were as follows:

Spanish Nationalists: Approximately eighteen Savoia-Marchetti S-79s and some Ju-52s, two wings of Fiat CR-32 *Chirris* and three wings of He-51, He-45 and Ro-37 fighters. The Spanish Nationalists could put about 106 machines into the air.

Condor Legion: Approximately seventy to eighty operational aircraft, between its fighter wing, composed of three squadrons of Messerschmitt Bf.109Bs and Cs at La Cenia, and its He-111 bomber wing at Sanjurjo (Zaragoza), as well as its reconnaissance squadron of Dornier-17s and He-70s.

Aviazione Legionaria: Three fighter wings with nine squadrons of Fiat CR-32 *Chirris*, and groups and independent squadrons of S-79 and S-81 bombers, fast Fiat BR-20 bombers and low-level attack Breda Ba- 65s in addition to the two squadrons of Ro-37 reconnaissance aircraft.

Spanish Republicans: Seven squadrons of I-16 *Moscas*, four squadrons of *Chatos*, together with four squadrons of *Katiuska* fast bombers and about seventy *Natacha*, Grumman and Vultee reconnaissance and low-level bombers. The number of aircraft per squadron varied between eight and thirteen.[4]

According to this account, the total of airworthy machines available at the beginning of the battle of the Ebro was as follows:

Spanish Nationalists, *Aviazione Legionaria* and Condor Legion	408
Republicans	250

Crude numbers, however, do not account for the characteristics and quality of aircraft, the training of the pilots and the tactical and strategic use made of the material available.

When the Republican forces had crossed the Ebro on 25 July 1938, all their available aircraft were occupied in holding back the Nationalist offensive towards Valencia. There was no Republican air force to cover the first week of fighting on the Ebro. One may ask why not, given that the Nationalist reaction ought to have been expected. This was a major error in the planning of the crossing of the Ebro. Without gaining control of the sky first, a force attacking over water risks its beachhead being destroyed by the defending air force.

In Catalonia itself, there was a squadron of out-of-date low-level *Rasantes* (though if they had been in Nationalist hands Franco's air force officers would have probably found an efficient way to use them as they did other out-of-date aircraft such as the He-51 German fighter biplanes). There were two bomber squadrons of *Natachas*, which could fly safely only with strong fighter escort, a squadron of Grummans and about a dozen *Chatos* and *Moscas*.[5] By 2 August, Republican air activity had increased. But German 88mm *flak* batteries downed a *Katiuska*, and on 13 August, Messerschmitt Bf-109s shot down six *Moscas*.

However, the Republican air force was receiving *Chatos* manufactured in factories in Catalonia and the Levante. Nevertheless, manufacturing problems and constant interruption of production because of enemy bombing raids meant that only four *Moscas* were delivered during the Ebro battle. In September, heavy bombing hindered the delivery of *Chatos*. Only three of these biplanes were produced in October.[6] Twelve *Supermoscas* came into service in August 1938, with new engines allowing the machines to operate at heights up to twenty-four thousand feet. However, at great height and low temperatures, the gun mechanisms froze and it took some time for the chief engineer to devise a method of passing hot exhaust gases around the guns. Now the *Supermoscas* were able to offer a difficult challenge to the Bf-109s, while they lost only two machines when on 15 August they downed twelve *Chirris.* On 20 August, *Supermoscas* downed two He-111 bombers and a Bf-109, and on 1 September the new planes downed three Bf-109s. On the same day four Republican pilots, Captains Leopoldo Morquillas, Juan Comas, Ladislao Duarte and Manuel Zarauza, most of whom were trained during the war, were promoted to the rank of major. But such was the dearth of pilots that the third cohort of pilots trained at Kirovabad, who had just returned but were not yet ready for active service, were sent up in early September with the consequence that some of them, flying *Chatos,* were soon shot down by CR-32 *Chirris*.[7]

Franco's German allies reacted swiftly to the Republican crossing of the river. General Volkmann immediately offered the full support of the Condor Legion, sending three 88mm batteries, and before midday on 25 July, barely twelve hours after the first units of the Republic had crossed the river, the Condor Legion's He-111 bomber group, the reconnaissance bombers and the *Stukas* had arrived to fly missions destroying the pontoons and footbridges. Over the next three months the bombers of the Condor Legion, which from 1 November 1938 the now General Wolfram von Richthofen was commanding, Volkmann having been recalled, would try day after day to hammer strong points, disrupt the enemy's communications and destroy bridges, which the Republicans repaired overnight.[8] Although the day-to-day German reports regularly mention anti-aircraft fire, this does not seem to have been effective. Few impacts were noted, though a Dornier-17 was shot down on 5 August.[9] Was this because Republican anti-aircraft fire was insufficiently concentrated, were too many shells dud or were the gun crews merely inefficient? Was the height at which the bombers flew (between 7,000 and 12,000 feet) too high for anti-aircraft accuracy?

The absence of any Republican effective countermeasures against bombing meant that it took much longer than the few days planned to build the metal bridge to allow the heavy artillery, the trucks and the tanks to cross the river. When García Lacalle, observing the battle from the Army of the Ebro's command post, suggested assembling all possible aircraft and bombing Gandesa, his plan, approved by Modesto, the commander of the Army of the Ebro, was overruled by the Staff in Valencia.[10] Was this treason merely unimaginative inefficiency or understandable unwillingness to risk losing aircraft?

Even when Republican aircraft appeared over the battle zone, the major complaint in memoirs against the Republican air force is its absence when most needed. Andrés García Lacalle writes frankly:

> Evidently our air force helped the land forces by bombing and machine-gunning the front, or protecting our men from enemy aircraft, but we were not able to achieve this aim with the efficiency, intensity and consistence that our soldiers needed. They would ask for our services urgently, and we tried to meet their requests as best we could, but once we had done what was needed, we had to get back to our usual tasks. Certainly there was less mutual understanding between air and land and thus less collaboration.[11]

What the 'usual tasks' (*nuestro normal desempeño*) were when the major battle of the Spanish Civil War was being so desperately fought is mysterious. The Nationalists, in contrast, concentrated their air forces on the Ebro. And why the links between Republican land and air forces were weaker than they should have been is not explained. Some opinion would blame the circumstance on the persistence of Soviet control over the Republican air force even as late as the battle of the Ebro, though why the Soviet advisers would want to deprive the communist Army of the Ebro of air support is hard to imagine. That the air staff may have been infiltrated by professional officers disloyal to the Republic is certainly possible. More likely, the cause is the lack of imaginative efficiency at highest level, a problem which to some extent had been remedied among the Nationalists.

The latter, well-armed with fast bombers, attacked poorly defended Republican aerodromes, destroying machines on the ground, while it was not easy to find experienced pilots for the fast *Katiuska* bombers to bomb enemy aerodromes, which were very well protected by the German 88mm anti-aircraft guns. Bombing of aerodromes and shortage of aviation fuel also hindered the Republican effort in the air.

The battle of the Ebro had become a war of attrition, of bombing, heavy shelling and positions taken and retaken by savage, hand-to-hand infantry combat. For Lieutenant Colonel Plocher, chief of staff of the Condor Legion, talking later to an American historian, 'On a small scale, it reminded one of the battles along the Western Front during the Great War.'[12]

By mid-August 1938, the Republican air force was fully engaged on the Ebro against the Condor Legion's Messerschmitt fighters and the highly trained military pilots of the Fiat CR-32 fighters with their powerful Breda 12.7mm guns. The particular advantages in dive, turn and general manoeuvre of the opposing fighters contrasted always with the inexperience and shorter training of most of the Republican pilots and the difficulty of replacing lost machines. Over twenty-eight days of combat, the Nationalist *La Cucaracha* wing downed twenty-three *Moscas* and fifteen *Chatos:* the *Asso di Bastone* wing shot down twelve *Moscas*, a *Chato* and two *Katiuskas*, and the *Gamba di Ferro* wing destroyed four *Moscas*. Over the entire battle of the Ebro, the *Aviazione Legionaria* shot down seventy-eight enemy aircraft, mostly *Moscas*, and many more were damaged, while the Condor Legion accounted for twenty-three and the Spanish Nationalists, with possible exaggeration, claimed ninety-eight. Fifty-nine Republican pilots were lost The Italians lost nine aircraft, the Condor Legion eight and the Spanish Nationalists twenty-four.[13] Whatever the exaggerations of claims by Nationalist pilots, it is evident that the Republican air force was close to collapse. Those days saw incessant flying by all the air forces involved and innumerable confrontations and not infrequent collisions. But the later Bf-109 versions were more than a match now for the I-16 *Moscas*, even possibly the *Supermoscas*, and German 88mm anti-aircraft guns shot down extremely vulnerable Russian SB-2 *Katiuska* bombers. Furthermore, Republican fighter pilots were reporting that the Condor Legion's He-111 bombers were so well armoured that even when peppered from close up by the four machine guns of the *Mosca*, the German bombers made it back to their bases.[14]

Unlike in the First World War, flying crew wore parachutes and often survived being shot down. In some cases, captured aviators were known personally from pre-war to their captors and were well-treated and exchanged. On 3 October 1938, for example, at the height of the battle of the Ebro, Captain Julio Salvador Díaz-Benjumea, later Lieutenant General and Minister for Air between 1969 and 1974 in one of Franco's governments, was shot down. Andrés García Lacalle, who had served under him as a sergeant in a squadron at the Tablada base near Seville in 1933–4, was invited to go to meet this highest-ranking prisoner at headquarters in Barcelona. An indication of the lack of military formality in the pre-war air force is that the ex-sergeant Lacalle greeted the ex-lieutenant Salvador at once with a salute and then addressed him as 'Salvador' and in the familiar form 'Tú'.[15] Lacalle assured his prisoner that he would be safe but that there might be some difficulty in effecting an exchange, given Salvador's high reputation as a fighter pilot. Lacalle took him on a drive around Barcelona in a chauffeur-driven car and then, after a beer and a meal, took him to the airfield at Valls. Lacalle wanted to persuade Salvador to communicate with his side to ensure that Republican pilots were also treated with respect when they were captured. Salvador promised to write to Joaquín Gómez Morato, commanding Nationalist fighters. Lacalle notes that there was consequently some improvement in Nationalist treatment of captured pilots, though the major problem was that they were kept in ordinary prisons and were not in the charge of the Nationalist air force. At Valls Lacalle allowed Salvador to climb into a *Mosca* and had the young pilots form up and introduced Salvador to them, during which time there was a raid and they had to take shelter. Salvador was eventually exchanged for a Republican pilot who had parachuted from his *Chato* in the northern zone in August 1937.

The Republicans were resisting fiercely. On 4 October, *Katiuska* bombers under their commander, Leocadio Mendiola, another ex-sergeant pilot now a lieutenant colonel, struck La Cenia, the main Legion base. However, Republican losses were serious. Such was the shortage of fighters that a pilot who deserted to Franco's side reported that Republican pilots were ordered to tackle enemy bombers but not their fighters. The outstanding Condor Legion fighter pilot Lieutenant Werner Mölders was achieving the kills which would make him one of the aces of the Spanish Civil War. Volunteering for Spanish service, he had arrived in Cádiz on 14 April 1938, flying first He-51s and then Bf-109s. When Adolph Galland was rotated back to Germany the following month, Mölders became squadron commander. Over the remaining months of the year, he became the leading ace of the Condor Legion, claiming fifteen 'kills': two *Chatos*, twelve *Moscas* and one *Katiuska*. Promoted to captain, he returned to Germany in December 1938. He was later killed in an air accident, having developed one of the great tactical innovations of war flying, known as the 'finger four', in which the aircraft flew in positions corresponding to the fingertips of an outstretched hand. This provided greater flexibility and mutual protection than the standard 'Vic' formation of the time.

The battle of Catalonia

On 23 December 1938, urgent messages summoned all Republican personnel on leave to return to their units. Franco's forces had crossed the Segre into Catalonia. No longer did the future Spanish dictator need to fear a hostile French reaction if his divisions reached the frontier, for France had accepted that Italian forces would remain in Spain until Franco's victory, and the latter had promised his neutrality in case of a European war. He had assembled overwhelming masses of men, aircraft and artillery against the exhausted Republican armies of the Ebro and the East. If his advance through Catalonia was slow, this was because of the bad weather as much as the sporadic attempts at resistance by Republican troops. By 6 January 1939, the Nationalists had occupied 1,600 square kilometres. In that first fortnight of the battle Republican communications were destroyed, while troops lacked supplies and were endlessly harried by bombing. As he prepared his forces for the final push through Catalonia, Franco could assemble 134 Italian-crewed machines and 146 flown by Spaniards, while the Condor Legion put ninety-six aircraft into the air, including some of the latest Bf-109s armed with larger-calibre cannon.[16] The Italian CTV and Franco's divisions advanced behind devastating fire from the Condor Legion's mobile 88mm guns followed by bombing and strafing aircraft. Throughout the first day, Christmas Eve, not a single Republican plane was seen, though infantry resisted fiercely at first. Such resistance, however, was futile in face of the Condor Legion's *Stukas* and the feared 88s. At the beginning of the new year the Republican air force recovered and became more active, but the newest and faster Bf-109s of type E, arriving from mid-December 1938 onwards, triumphed over the inexperienced Republican pilots. The Condor Legion moved its command posts forward frequently. German aircraft bombed Republican aerodromes with impunity, destroying many machines on the ground, largely because of weak anti-aircraft defences, although the Barcelona port area was strongly defended and the *Stukas* were pulled out.[17] The rest of the battle of Catalonia was a walkover. Huge numbers of prisoners were taken. In Barcelona, morale plunged. Refugees fled towards the French frontier. Among the Republican leaders there was doubt about whether to continue to resist in the other part of the Republic, known as the Central-South Zone, a triangle with its angles at Madrid, Valencia and Cartagena. Catalan industrial towns fell one by one, and by 24 January Franco's armies had reached the mouth of the Llobregat just south of Barcelona. On 26 January 1939, this great city was occupied without resistance by the Nationalists. Republican troops, accompanied by tens of thousands of civilians and the Republic's leaders, fled towards the frontier. All attempts to create last-minute centres of resistance failed. On 4 February, Gerona was occupied and the following day the President of the Republic, Manuel Azaña, the Prime Minister, Juan Negrin, together with the Presidents of what had been the autonomous regions of Catalonia, Lluis Companys, and of the Basque region, José María de Aguirre, crossed into France. Between 5 and 9 February 1939, the armies of the East and the Ebro were admitted to France and interned in grim, cold and rainy camps on the wintry beaches of French Catalonia.

In the ten days after the fall of Barcelona, Republican aircraft often found difficulty in landing after operations, so fast was the enemy's advance. Many were destroyed on the ground. The actual number of machines was falling fast and lost aircraft were not being replaced. Sixty-nine *Moscas* were captured in various stages of construction and testing.[18]

Had Stalin abandoned Spain? It would have seemed so but for the Russian leader's extraordinary response to a final appeal from the Republican leader, Juan Negrín. Some days before the great Franco offensive against Catalonia, Negrín ordered the commander of the Republican air Force, now a general, Ignacio Hidalgo de Cisneros, to leave immediately for Moscow to appeal personally to Stalin for a massive shipment of tanks, aircraft, artillery and machine guns. A day after his arrival he was seen by Stalin, Molotov, the Foreign Minister, and Marshal Voroshilov, the army chief. Hidalgo de Cisneros was amazed that Stalin agreed to the demands which he somewhat embarrassedly presented for material costing 103 million dollars, given that the reserves of gold which had been shipped to the USSR in October 1936 were by now exhausted. Stalin said that Hidalgo de Cisneros should discuss the matter of payment with the Commerce Minister, and then invited his Spanish guest to bring his wife, Constancia de la Mora, and to have supper with the three powerful Soviet statesmen.[19]

Seven ships set sail almost at once, from Murmansk for Bordeaux and Le Havre with 104 aircraft on board as well as much other weaponry, but difficulties placed by French frontier officials meant that very few entered Spain. García Lacalle, at the time commander of the entire Republican fighter arm, recalls that a ship arrived with thirty I-15B *Chatos* of an advanced specification. They were assembled and crewed in the last days of January 1939, but the enemy's advance meant that they had to be moved at once to Figueras, close by the French frontier. Here all the aircraft remaining in Catalonia were concentrated in one large aerodrome, where the officers strove to achieve some sort of coordination. Lorries, troops, government employees and vehicles crowded the streets, under constant enemy harassment. *Chatos* and *Moscas* found it hard even to take off, because the CR-32 *Chirris* were waiting to attack before the Republican fighters could reach their ideal combat height. Some of the Republican squadrons had to be moved to the neighbouring aerodrome of Vilajuiga, whose air-raid shelters were overrun by civilian refugees every time there was an air raid. Republican aircraft were destroyed on the ground, and so some *Chatos* and *Moscas*, finding themselves unable to land because the aerodromes were full of craters, flew to France. On the nights of 2 and 3 February 1939, García Lacalle heard cartridges exploding during the night as, without unloading their guns, the mechanics blew up all planes which required more than forty-eight hours to be repaired.[20] At about the same time the Republican pilot, Juan Lario Sánchez, saw a goods train stopped at the frontier town of Port Bou with its engine steaming, at the head of long line of flatcars with aircraft on them which were evidently not being allowed to cross the frontier. On 8 February, he and his comrades walked for hours over the Pyrenees into France, where the beach camps of Argelès and St. Cyprien-Plage awaited them.[21] García Lacalle recalled seeing large packing cases abandoned in open country, containing wings or fuselages of Russian fighters and bombers. By this time, he writes, 'Why did I want them, when I did not even have suitable pilots to fly them?' His veteran pilots were exhausted, his new ones inexperienced. He sent the cases back over the frontier.[22] On 5 February, Lacalle ordered twenty-nine recently arrived I-152 *Supermoscas* to be flown into France. On the night of 5–6 February 1939, Vilajuiga was practically full of *Natachas*, *Rasantes*, *Chatos*, *Moscas* and others, forty-three in all. The next morning García Lacalle and some others flew into France, pursued by Messerschmitt fighters. He landed at Toulouse, from where the first Dewoitines to go to Republican Spain had taken off in August 1936. The other aircraft at Vilajuiga were destroyed on the ground by Messerschmitts.

Leaving aside a briefly successful Republican campaign in Extremadura in early January, this was the end of the Spanish Civil War. In the Central-South zone there remained a couple of squadrons of *Katiuska* bombers, three squadrons of *Natachas*, now completely out of date, and some thirty I-16 *Moscas*. Unsurprisingly, Colonel Camacho, now commander of what was left of the Republican air force, told Prime Minister Negrín, at a meeting of Republican military leaders held at Los Llanos aerodrome near Albacete on 16 February 1939, that there was no point in resisting Franco.[23] When on 27 February Britain and France gave *de facto* recognition to the Franco government, the President of the Republic, Manuel Azaña, refused to return to the Central-South zone, thus removing legitimate authority from Negrín, who had flown back to Republican Spain and tried to encourage resistance, against the advice of almost all the military leaders. Consequently, a long-planned uprising by career officers, led by Colonel Segismundo Casado, commander of the Army of the Centre, overthrew the Negrín government, overcame resistance from the communist-led divisions in the zone, who wanted to resist to the end in order to allow the evacuation of people who faced the greatest danger from Franco's execution squads, and attempted to reach a settlement with Franco, who insisted on nothing less than unconditional surrender. At the end of March, the Republican armies of the Centre, of the Levante, of Andalusia and of Extremadura, melted away as Franco's forces occupied the rest of the Republican zone. On 1 April 1939, the victorious generalissimo announced that the war had ended.

Meanwhile, aircraft played a significant role in peace negotiations and in evacuating Republican leaders. Once the victory of the Casado coup was assured, the political leaders of the Republic who were still in the Central-South zone, as well as the Central Committee of the Communist Party and the communist militia leaders, feared for their lives. For their evacuation, Negrín ordered Colonel Camacho to send two DC-2s from the latter's headquarters at Albacete to Monóvar, a small airstrip close to where Negrín had established his offices in Elda (province of Alicante). 'Let's go!' (*¡Vámonos ya!*), Negrín is reported to have said, given that every possibility of resistance to Casado's coup and the surrender to the Nationalists which would doubtless follow was hopeless. The aircraft were ready: two DC-2s and a De Havilland Dragon Rapide, like the one which had taken Franco from the Canaries to Spanish Morocco two years and ten months earlier. At 3 pm on 7 March 1939 the first DC-2 took off for Toulouse, followed half an hour later by the second. There was no room for the last Soviet adviser, General Shumilov, who was not in danger of his life and returned to the Soviet Union in May 1939. The Dragon Rapide had flown to Oran carrying the communist leader Dolores Ibarruri (a. 'La Pasionaria'), General Cordón, the Undersecretary for War, and two or three others.[24] All was over. All that remained was for Colonel Casado's National Defence Council to arrange the best terms possible with Franco.[25]

On 23 March 1939 another of the DC-2s of the Spanish civil line LAPE took off from Barajas carrying two senior Republican staff officers, representing the Council of National Defence (*Consejo de Defensa Nacional*), which Colonel Casado had established with the aim of ending the war. They flew to Gamonal aerodrome outside Burgos, where they conferred with Nationalist officers. As for aircraft, the terms for surrender were that on 25 March all machines should be surrendered with their armament. Some Republican pilots took off without orders, flying to Oran in French Algeria, whence fifty-eight machines, among them three *Katiuskas*, four *Chatos*, thirteen *Natachas* and sundry other warplanes and civil aircraft, were later recovered by the Spanish Nationalists after agreement with France.[26] On learning about these flights, when the Republican colonels returned on 25 March for further negotiations about surrender, the Nationalist representatives suspended talks and ordered the Republican representatives back to Madrid. Between 26 and 31 March, Franco's armies overran the Central-South zone, with its capital, Madrid, and the coastal cities of Valencia, Alicante, Murcia and Cartagena, and the provincial capital of Albacete, which was the headquarters of the Republican air force. Now the flight into exile began, when thirteen military aviators boarded a civil aircraft, used to train observers, at Los Alcázares (Murcia) and flew it to Oran. In another case, fighter pilots agreed to surrender rather than fly to Algeria because the ground crews feared they would be subject to reprisal if they supplied them with fuel. On 29 March, most of the pilots of the Republican air force stationed in the zone flew the remaining machines to Barajas. Some landed on the way, at the aerodrome of Los Llanos in Albacete province, where Colonels Antonio Camacho and Manuel Cascón confirmed the orders to hand their machines over to the victors. At Los Llanos there were three *Katiuskas* under the orders of Lieutenant Colonel Mendiola, who took Camacho into exile, although Colonel Cascón surrendered to the Nationalists, to be charged with military rebellion and executed. On 28 March, as Colonel Casado flew from Madrid to Valencia, beneath him he saw soldiers making their way home.[27]

What happened to the losers?

Rafael Ballester Linares was a second-year chemistry student, who flew for the Republic as an observer in a *Katiuska* bomber. On 16 November 1938 his squadron took part in the bombing of the Condor Legion's main aerodrome at La Cenia just on the southern border of Catalonia, flying from Figueras under the command of Lieutenant Colonel Leocadio Mendiola. They destroyed two German fighters, but two Messerschmitt Bf-109s, possibly type E-1s, just arrived from Germany, equipped with more powerful engines, caught up with the *Katiuskas* and shot them down with a few blasts from their 20mm cannon.[28] Ballester's *Katiuska* was hit by anti-aircraft fire and crashed. He baled out. Arrested by the local *Guardia Civil*, he was well-treated, fed by the wife of the guard and given a pair of *alpargatas* (espadrilles) because his flying boots had lost their soles. The guard even forgot to search him and remove his revolver. In time, a car came to take him away. He found himself among a group of Germans who took him to their base, which he had taken off earlier to bomb. They interrogated him at length, boasting about how good German

Intelligence was on the subject of Republican air operations and the location of many 'secret' air strips. Over frankfurter sausages and bottles of beer, the Germans assured him that he would be exchanged. Jokingly, the Germans told Ballester that his raid had destroyed their brothel. When Ballester said that there were no field brothels in the Republic air force, the Germans assured him that this was a serious mistake. They were the only men who had access to the girls and their doctor examined them regularly, so the Germans ran no risk of disease. Then he was taken to Zaragoza. On the way they stopped at an aerodrome and, at 2.30 am, woke up the pilots and spent a riotous time in the bar. The Spanish airmen said they would try to keep Ballester with them until an exchange could be arranged, though this was unlikely to happen because the senior officers were pigs (*marranos*). While at Zaragoza, his interrogator asked him in all seriousness, how he, an educated and intelligent person, could have served the Reds. Next, he was driven to prison at Salamanca and put into a room with other captured pilots. At Christmas, they received numerous parcels from the Red Cross.

After this remarkably good treatment, one would have thought that Ballester would have been exchanged or at least released at the end of the war. On the contrary, when he was tried for 'military rebellion' he was sentenced to death, commuted to twenty years and one day. He was not released until 1944.[29] This would be the experience of many pilots of the Republic who did not leave Spain. The senior officers, Antonio Pastor, Antonio Camacho, Felipe Díaz Sandino, Carlos Núñez Maza, Ignacio Hidalgo de Cisneros, Leocadio Mendiola, Andrés García Lacalle and others, who would most likely have been executed, managed to leave Spain. Some never returned. Others, among them Leocadio Mendiola who commanded the bombers, came back to Spain after 1966, when a general amnesty was announced for 'crimes' committed in the Civil War. Nevertheless, the list of court martials as published by the Spanish air force long after the death of Franco in 1975 quotes hardly any death sentences.[30] But, since it does not include the death sentence passed on Colonel Cascón, who surrendered the Republican air force at the end of the war, the list, which includes 3,260 names of men of all ranks, may be incomplete. Some examples of sentences include the one handed down to Isidro Giménez García, a junior officer in 1936 and later commander of the Republic's fighter squadrons. He was sentenced to twelve years' imprisonment, as were Angel Riaño, a senior career officer who served as chief of staff, and Juan Comas, a naval pilot in 1936, who commanded a wing of *Chato* fighters. Whether they were condemned to shorter or longer terms of imprisonment, or benefited, as most did, from amnesties and conditional releases, every man who received a sentence of over three years was dismissed from the service, lost all rights to pension, which were not restored until many decades later, and had to make his way and support his family in the atmosphere of hostility that ex-Republican military men endured for years after the war.

Most of the other leaders of the Republican air force ended up in Mexico, where some had very successful careers. Among them was a young pilot, Manuel Montilla, who had taken the course in the USSR and crossed into France in February 1939, and managed to reach Mexico. Here, in partnership with Leocadio Mendiola, he founded an air cargo company. He returned clandestinely to Spain in 1969 to organize an illegal political party. Finally, his rank as colonel was recognized by a post-Franco government. He died in Spain in May 2007.[31] Antonio Camacho, air force colonel, Undersecretary for Air and commander in the Central-South zone, lived modestly in Mexico as a sales representative, while Angel Pastor died in an old-age home in France, denied the return to Spain which he had requested.

Others spent most or all of the rest of their lives in the Soviet Union. Ignacio Hidalgo de Cisneros, the only convinced communist among them and a member of the Party's Central Committee, left Spain on 6 March 1939 piloting a DC-2 carrying Negrín, the communist leader Vicente Uribe and the communist militia officers Manuel Tagüeña and Enrique Líster. He went to the USSR, but left, went to Mexico and became honorary president of the Association of ex-Spanish Republican Pilots. After the Second World War, he lived in Poland, and later in Prague, dying suddenly in 1966 in Bucharest.[32] Another group of younger pilots who were in Russia when the Spanish War ended, flew for the Soviet Union during the Second World War. Juan Lario Sánchez, for instance, an apprentice in an architectural drawing office in 1936, was selected for pilot training. He was a member of the cohort which spent six months at Kirobavad. He crossed the frontier into France on the defeat in Catalonia in1939 and was selected to go in a group to the USSR. Throughout the Second World War, he flew against the Luftwaffe. No till 1967 did he return to Spain. Leopoldo Morquillas, a corporal on leave in Madrid in 1936, became a pilot, led a squadron and was selected to attend a course in tactics in the Soviet Union. During the Second World War, he led a Soviet squadron and was later a divisional inspector. Retiring in 1948, he became a factory manager at Tula, where he lived for the rest of his life.

Of the Soviet airmen, Yakov Smushkevich (alias 'General Douglas'), commander of the Soviet air force in Spain from November 1936 until May 1937, was executed in October 1941, probably as a result of the poor showing of the air force against the German invasion. He was followed in Spain by Vsevelod Lopatin (alias 'General Montenegro') from May to September 1937 when he was withdrawn, possibly because of strains with the Spanish Air Minister Prieto. Arrested and accused of anti-Soviet conspiracy, Lopatin was executed on 29 July 1938. He was replaced by Evgenii Ptukhin (alias 'General José'), who was arrested and executed, probably as a result of the collapse of Soviet forces following the German invasion of 1941. The last Soviet chief in the Republican air force was Alexandr Andreyev. Other Soviet aviators in Spain, however, went on to enjoy distinguished careers in the Soviet air force.[33]

Of Franco's pilots, Julio Salvador Díaz-Benjumea and Eduardo González Gallarza became senior generals and ministers in Franco's government, while Angel Salas Larrazábal volunteered for service against the Soviet Union in the Second World War, making seventy operational sorties and destroying sixteen enemy aircraft. He reached the rank of lieutenant general, retiring in 1972. The greatest Spanish ace, Joaquín García Morato, who achieved forty credited victories over Republican aircraft, died shortly after the war, on 4 April 1939, when his Fiat CR-32 crashed while he was performing low acrobatics for newsreel cameras.

As the Italian fighters soared over the heads of the spectators of the victory parade in Madrid, their pilots thought perhaps of the aviators that the Italian air force had lost in the Spanish war.[34] On 12 May 1939, all the almost five hundred machines in Franco's service were reviewed by him at Barajas, today's civil airport near Madrid. A total of 147 Italian and eighty German aircraft were left in Spain to equip a semi-modern Spanish air force. On 10 June, 1,800 Italian aircrew and ground staff of the *Aviazione Legionaria*, accompanied by General Kindelán, Franco's overall air force chief, embarked for the voyage home from Cádiz on the liner *Duilio*, which called at Palma to pick up the flyers of the *Aviazione delle Baleari*. Italy had sent altogether more than seven hundred aircraft and six thousand aircrew to Spain. A total of 193 Italian airmen had lost their lives, while 86 Italian aircraft had been lost in action and one hundred in accidents. Many of the younger Italian pilots would be among the 3,007 who lost their lives in the Second World War. Their commanders, Velardi, Bernasconi and Monti, went on to become senior generals. After the September 1943 armistice, Major Duilio Fanale sided with the Allies and went on to fight against Germany. Colonel Ruggero Bonomi, on the other hand, who had led the twelve bombers in the pioneer flight from Sardinia to Morocco on 30 July 1936, was a major general by the time of the Italian surrender in 1943. Loyal to Mussolini, he sided with the Repubblica Sociale Italiana and after the war was dismissed from the air force.

The Condor Legion bids farewell to Spain

On 6 February 1939, waves of German planes bombed and strafed the frontier town of Figueras, choked with fleeing troops and civilian refugees. This would be the final mission of the Condor Legion, although it now moved its headquarters to Toledo, in the Central-South zone, in preparation for the offensive that was intended to overrun remaining Republican territory. At 10 am on 27 March, however, following the Republican surrender, von Richthofen ordered all operations to cease. German aircraft flew in mass formation over Madrid, followed by several other similar events in other cities before the final parade on 22 May at León and the departure of the Condor Legion from Vigo on 28 May 1939. Hermann Goering, overall head of the Luftwaffe, met them when they docked at Hamburg. They were transferred to Döberitz, their original assembly camp, before returning to their units. On 6 June, Hitler reviewed a massive and triumphal parade of the whole Condor Legion, that is all the approximately twenty thousand men who had gone to Spain save the 298 who had lost their lives.

As for the pilots of the Condor Legion, just as the secret German air force of the 1920s analysed and reanalysed the lessons of the First World War, so those who had flown in Spain wrote and studied their and others' experiences in order to draw the appropriate lessons, a process which went on till the end of the first year of the Second World War.[35]

The success of Wolfram von Richthofen in Spain, and the triumph in 1939–40 of the application in Poland, Norway and the Low Countries of the techniques learnt in Spain, led to his promotion to the exalted rank of Field Marshal in February 1943. Nevertheless, a large number of the surviving German fighter and bomber pilots who had flown in Spain met their deaths in the Second World War. Almost all had been promoted to lead squadrons. Adolf Galland survived the war and rose to general rank; Mölders was killed in a flying accident, Von Bonnin died in Russia, Oesau in Belgium in 1944 and Lützow in 1945 over Germany. A small number who reached high rank and served in rearguard positions and the older officers who had fought in Spain such as Harlinghausen, commander of seaplanes flying out of Majorca and Plocher who had been chief of staff of the Condor Legion, survived as high-ranking staff officers. Sperrle was acquitted of war crimes by the 1946 Nuremberg tribunal, while von Richthofen died of a brain tumour after the war and Volkmann in a motor accident.[36]

In the post-war German army, the *Bundeswehr*, a number of ex-Condor Legion men such as Hans Henning von Beust, who had been a colonel in Spain, and the ace Hannes Trautloft, who had learnt to adapt themselves to the new democratic Germany, finished their careers as generals.[37]

It was, understandably, the issue of Guernica which had to be settled before an end could be put to the inherently strained relations between post-unification Germany and post-Franco Spain. Versions published in Spain at the end of the Franco regime slowly began to accept that the Condor Legion had carried out the bombing of Guernica, though the question of the origin of the orders to do so remains unclear even now. For pro-Francoists, the Germans bombed the town without Franco's orders or consent. But ex-Condor legionaries objected to being linked closely to the cruelty of Nazism. In 2014 the President of Germany, Roman Herzog, publicly admitted the 'culpable involvement of German pilots in the bombing'.[38]

Chapter 10
Conclusions

Flights marked the beginning and the end of the Spanish Civil War. A plane crash killed General Sanjurjo, the intended ruler of the new Spain to arise from the insurrection of July 1936, while a second air catastrophe killed General Mola, a possible rival to Franco. In contrast, a successful long-distance flight by a group of British people and a Spanish journalist from Croydon to Las Palmas and thence to Spanish Morocco put Franco in a position to take over the leadership of the military insurrection against the Spanish Republic. And at the end of the war, some of the leaders of the Spanish Republic would escape by flying out of Spain to France and Algeria.

It was the chance occurrence in the week after the army insurrection in Spanish Morocco of the availability of a long-distance German aircraft to take Franco's emissaries to see Hitler and obtain military aid that enabled Franco to bring his professional army to the Spanish mainland from Morocco despite the Republican naval blockade of the Strait of Gibraltar. The Junkers-52 machines which Germany sent to Franco were intended as transport planes rather than bombers, and the Heinkel-51s which accompanied them were for defence, but they were the precursor to the Condor Legion itself.

Italian bombers dispatched to Spanish Morocco on 30 July 1936 heralded the major supply of over 750 Italian aircraft that would give Franco overwhelming air superiority.

Germans and Italians brought modern aircraft to the Spanish war, which began with aircraft designed in the 1920s and with a style of individual combat which would have been immediately recognizable to a the First World War pilot. Navigation, bomb-aiming and communications, for example, among other advances in air warfare, were undeveloped in July 1936 when the Spanish Civil War began. When it ended two years and eight months later, air battles were being fought with fast monoplane bombers and fighters of types which would be flown in the Second World War.

At every stage in the war the dominance of the skies by Franco's Nationalists gave them the edge over their opponents, from the beginning when a combination of strategic confusion in the Republican command caused the latter to fritter away its larger number of aircraft instead of identifying and concentrating on strategic essentials such as blockading the Strait of Gibraltar and supporting the attempt to retake the island of Majorca for the Republic. At the same time, most of the pilots with greater flying time in their log books were commissioned officers sympathetic to Franco. Hundreds of air force officers stationed in what became the Republican zone were dismissed because, rightly or wrongly, their loyalty to the Republic was distrusted. Thus, the Republican air force had to rely on sergeant pilots with far less flying experience and very soon on massive training programmes for new aircrew.

The immediate result of German and Italian aid to Franco and of Republican error was a highly successful airlift of troops and equipment in the early weeks of the war from Morocco over the Strait of Gibraltar and a virtual Italian takeover of what would become a vitally significant naval and air base on Majorca. Later on, Franco's aircraft, even in very small numbers, were able to help his relatively efficient ground forces put to flight the inexperienced militias of the Republic. Later still, the combination of German efficiency and constant practice in close air–ground cooperation, together with the technical superiority of the Italian Fiat CR-32 fighter, the *Chirri*, practically drove the Republican air force from the sky and led to the Soviet Union sending important numbers of its at the time highly modern bombers and fighters to succour the Republic.

Between November 1936 and March 1937, Soviet fighters and bombers ruled the skies over central Spain, preventing Franco taking Madrid and putting a rapid end to the war. The German reaction was to put their aid on a firmer basis in the form of the Condor Legion. Messerschmitt Bf-109 fighters, in constantly updated versions, gradually overcame the Soviet threat, while later German bombers such as the Heinkel-111 and the Italian Savoia-Marchetti-79 were used efficiently to destroy factories and cities in the Republican rear.

Shipping difficulties and Moscow's internal conflicts as well as its hesitation, in the context of its political need to reassure the Democracies of its intentions, meant that Soviet aircraft were less readily replaced than Italian and German machines, and not in high-specification models until too late in the war. As for aircrew, while German and Italian professional military aviators were regularly rotated, Soviet pilots were recalled without enough and sufficiently trained Spanish crew to replace them.

Consequently, Republican pilots in the last months of the war were too young, too inexperienced, overworked, demoralized and possibly not sufficiently skilled to take advantage of the greater manoeuvrability of the Soviet *Supermoscas*, which might, if flown by experienced pilots, have been a match for the latest Messerschmitt Bf-109s. Some of the Republican pilots, especially in the midst of battles, might at certain times be in the air and suffering enormous stresses for up to seventy hours per month, while Condor Legion pilots would fly three or four missions a day at the most but often far fewer and followed by several days' rest.

German, Italian and Soviet experiences in Spain

While in theory the presence of the Condor Legion was secret, in practice it had been known from the beginning. In Germany itself, relatives of the men who had gone to Spain were strictly warned about spreading rumours that their sons and brothers were there, but enough of the apparatus of the German Socialist Party had survived the Nazi takeover for information to be spread to sympathizers abroad.[1] In Spain, of course, their presence could not be hidden, especially when at the beginning of 1937 the Condor legionaries received their Spanish uniforms which looked as if they had been 'cut by a Prussian tailor'.[2] Unlike the Russians, whose orders were to keep a very low profile in Republican Spain, and who in any case were far fewer in number, the Germans were very popular and applauded in the streets of towns in the Nationalist zone, in particular Avila, Salamanca, Burgos and Vitoria. Sometimes a German would drink in a bar and find that a stranger had paid the bill. German bands often entertained the public, while parades singing German marching songs, flags, swastikas and flying exhibitions were frequent.

The Germans took over public buildings. In Avila, for example, they were billeted in the social club or casino, in churches, schools and the telephone exchange. The senior officers were put up in the Hotel Continental, opposite the cathedral. That many of the Germans were Protestants did not seem to trouble the ultra-Catholic Francoists, who raised no objection when Condor Legion men built a chapel for themselves inside the monastery of St. Thomas, where many were housed and from which the monks had been removed. Sanitary facilities in the monastery were poor, so the Germans built a shower block. When the monks returned after the war, they condemned this elementary hygienic facility as 'sinful' (*pecaminoso*) and destroyed it.

While Adolf Galland, the later ace German pilot of the battle of Britain, was at Avila, he used to take time off, stay at the luxurious Parador de Gredos, the first of Spain's top-class state hotels, and hunt game with some of his comrades. Generally speaking, in the conservative and Catholic cities of Franco Spain, German officers were entertained and frequently accommodated in the prosperous homes of the bourgeoisie. Food was abundant, unlike in the Republican zone which had less fertile agricultural land. Whether Spanish cooking suited German tastes was another matter. The indispensable beer and sausages were supplied from Germany, but German stomachs were often upset by the ubiquitous olive oil in which food was cooked, though just as probably by overeating unfamiliar dishes, such as when, to celebrate the occupation of Santander in August 1937, the mayor of Burgos, the Castilian city where Franco's government was based, gave a reception in a luxury hotel, where German staff officers and senior officers enjoyed caviar, paté de foie gras, river crabs and ham. The following extract from the memoirs of Karl Weller, an International Brigades prisoner who had been handed over to the Condor Legion, gives a picture of the pleasant time that off-duty German officers enjoyed in Franco's Spain.

> On the table were full carafes of wine and dishes of fruit, as well as steaming tureens of soup, and a group of dark-haired young ladies dressed in white lace aprons coquettishly served the officers roast meat and other things on china plates.[3]

'Coquettishly' was not the right word to use. The Germans had to learn that, in Franco's extremely Catholic Spain, the friendliness, even the apparent flirtiness of Spanish women, was not sexual provocativeness. Indeed, even in 'Red Spain' female amiability was not intended to imply sexual availability, as the British hero of the 1995 Ken Loach film, *Land and Freedom*, discovers to his dismay in the Spanish militia trenches. The admiring glances of Spanish women at smart, fit, bronzed young Germans went no further than that. Indeed, some Germans were puzzled that they could not take a girl to a public dance and learned that in Spain dance halls were places where men went to find prostitutes. And when, one hot Sunday in the western city of Cáceres, German airmen swaggered around dressed in their best white shorts, the local girls looked away, embarrassed, as if the men had forgotten to put on their trousers.

In the small Castilian city of Avila there was a small aerodrome which had been laid out for a flying show in 1915 and had been used occasionally as a landing strip for pilots in difficulties. A hangar was swiftly built for the Condor Legion and the field was levelled so that within a few weeks almost all planes could land on it. Given the proximity of Avila to the front around Madrid and the good visibility of the clear skies of the Castilian plateau, the aerodrome became extremely busy. But after the war, pilgrims walking to the shrine of the *Virgen de Sonsoles* passed the airfield and were shocked to see the sexually explicit pictures that the Germans had painted on the hangar walls.

Nowhere were the particular needs of young men better catered for than in the brothels of Seville's large red-light quarter, from where the Condor Legion took a travelling group of about twenty girls when they went north to Vitoria. German organization required soldiers' and officers' brothels to be controlled by military police, and the women had to be medically examined. One English-educated Spanish insurgent pilot from the wealthy Larios family of Málaga recalled the Germans flooding into the Alameda de Hércules, Seville's red-light district.

> They [the Germans] ran these primitive joints on a rough military basis with the same thoroughness they would have run a barracks. The men were marched up in formation, and if by any chance the houses were full they would be lined up in the street ready to advance in single file, waiting patiently for orders from who[ever] was in command inside, who was probably clocking [timing] them with military precision. The whole setup seemed comic and out of place in Seville, and I could hardly imagine Spanish troops carrying out these 'duties' in this manner with straight faces. I often wondered what the veteran prostitutes had to say about this love regimented by the stentorian commands of a sergeant- major.[4]

On 1 December 1936, the chief of staff of the Condor Legion noted with dismay that there had been twenty-nine cases of sexually transmitted disease since his arrival a month or so before.[5] It seems a remarkably small figure.

'Sunny Spain' was rarely the gentle and warm country of northern imagination. In the Castilian winter, the ground staffs sleeping in the hangars struggled to find shelter from the icy wind.[6] This was a huge contrast with Seville and the plains around Madrid where summers are unbearably hot. Here the ground crews of the Condor Legion toiled almost naked and were bothered by flies and bitten by mosquitoes as they slept. Fighter pilot Harro Harder noted that, when flying over Brunete, in July 1937, the descent from a cool nine thousand feet to near ground level where the temperature was 46 degrees centigrade in the shade was like opening an oven door. But in the winter, on the Castilian *meseta*, only a few miles north of Madrid, temperatures resemble those of Poland and fall to many degrees below zero. It was worst in Teruel, in Aragon, where the battle of the winter of 1937–8 was fought in sub-zero Russian-like conditions.

For some Germans of the Condor Legion, Spain, especially the backward-looking Catholic Spain which they had come to defend, was centuries out of date, a different world. Von Richthofen's comment in a letter to his wife was, happily for German–Spanish cooperation, private. The chief of staff of the Condor Legion wrote, 'Life, surroundings, food, people, the country: all repugnant' (*Leben, Umgebung, Essen, Leute, Land, alles scheusslich*).[7]

Despite the embraces, exchanges of decorations and backslapping bonhomie on official occasions, there was inevitably tension between Germans and Spaniards. At a very profound level, the Germans were intensely irritated by what they saw as Spanish arrogant indiscipline, refusal to be hurried and insistence on their very late eating hours, when even senior officers put lunch and sometimes church attendance before punctuality. Orders were given from the top, but decisions on action seemed to be taken lower down by personal agreement. Von Richthofen compared Spanish written orders to children's planning of cops and robbers games.[8] The Spanish Civil War was, after all, a universe in which the arguably most zealous and efficient military officers in the world were cooperating with an army and air force which had not fought the First World War, whose experience was limited to colonial warfare and whose ways had hardly changed in a century. Henning von Beust, a German bomber pilot and later staff officer, referred in a report to Spanish 'misunderstanding and obstinacy'.[9] For their part, Germans, in particular General Sperrle, were not renowned for their tact, particularly when criticizing Franco's apparent lack of urgency and Spanish insistence on taking their time when the Germans wanted to be up and doing. Spanish elegance of manners and lordly refusal to be hurried clashed with German modes of prompt reaction and attention to the task in hand. Spaniards of course respected German professionalism, but squadron leader and Second World War ace Günther Lützow's diary complains that the Germans behaved in public with 'sickening vulgarity and arrogance'. Off duty, they often got drunk, which is despised in Spain.[10]

As for the enemy, the Republican 'Reds', the Germans came to respect them and even to think that their values, in so far as they reflected the 'socialist' part of National Socialism, were more to be admired than those of the wealthy, landowning and church- and army-dominated Franco zone. Upper-class Spanish gentlemen in their clubs and in church might well share hatred of communism with the Nazis, but they were alienated by the latter's lack of piety and their obsession with racial and physical values. Colonel Erwin Jaenecke, chief of staff of *Sonderstab* W, for example, after a visit to Spain, wrote in an official report that the Reds were fighting for a principle, but that it was doubtful if this could be said of the general population in the Franco zone.[11] It was common to hear pilots saying, 'We are fighting on the wrong side.'[12] And when, during the Second World War, Hitler was told that Franco had awarded the full honours of a captain general to the patron saint of Segovia, in recognition of the miraculous protection of the city during the civil war, the German leader declared that he would never ever go to Spain.[13]

As for the views about other peoples whom the Germans met in Spain, they were, despite their sense of racial superiority, particularly fascinated by the Moroccan troops in Franco's army, by their foolhardy courage and their savagery. Moroccan tourist articles, such as cushions and brass work, were popular gifts that Condor legionaries sent home.

German attitudes towards their Italian allies were mixed. While they admired the quality of the Italian aircraft, they tended to see many of the Italian pilots as boastful and arrogant, as if their victory against primitive forces in Abyssinia just prior to the Spanish war was in any sense admirable.

Many Spaniards, however, were much impressed by the swift Italian conquest of Abyssinia compared with the long-drawn-out wars of the 1920s in Spanish Morocco, as well as by Italian panache, high-quality equipment and even musicality. Like German officers, Italians were welcome in upper-class houses in the Franco zone. Their Latinity and Catholicism made them seem less unfamiliar.

For some Spaniards, however, as for the Germans, the Italian disaster at Guadalajara inspired a sense of *schadenfreude*. Italian lack of precision in their bombing technique was frequently commented on. Arrogance and theatricality were what most struck the perceptive American journalist Virginia Cowles. Italian officers strode 'booted and spurred', usually with a girl on each arm, in the Gran Hotel of Salamanca, and 'any girl who hasn't got a face like a boot, can get a ride in an Italian truck', grumbled the unpleasant Captain Aguilera, in charge of the foreign Press.[14] Given that Italians could make themselves understood more easily than Germans, and that they were Catholics, it is not surprising that even in very conservative Franco Spain girls fell for glamorous and attentive Italian officers. One sad case concerned an Italian lieutenant who was killed in action while his Spanish girlfriend was expecting their child. The girl's family refused to recognize the infant who was taken to his late father's family in Italy and later entered the diplomatic service. There were a number of marriages, however. Among others, Major Duilio Fanali married a Spanish nurse. Major Andrea Zotti, who led the XXIII fighter wing, married the daughter of General Kindelán, the commander of the Nationalist air force.[15]

Turning to the highly paid volunteer foreign pilots flying for the Republic, if they were based at airfields around Madrid, they were accommodated at the capital's top-class ten-storey Hotel Florida. Here, in the Plaza de Callao at the top of the Gran Vía, Frank Tinker, the American volunteer flyer for the Republic, once shared a small lift with the bulky Ernest Hemingway, who was sending dispatches to the North American Newspaper Alliance and was gathering material for his famous novel about the Spanish Civil War, *For Whom the Bell Tolls*. When the Florida's famously efficient hot water system was destroyed by a Nationalist shell, the pilots abandoned its dangerous rooms, which were in range of even small-arms fire. They were now put up in the basement of the Prado art gallery and slept under statues of Greek gods.[16]

While in Madrid, Tinker met a girl called Dolores and they 'kept company' during his free time, going to the cinema and having lunch. While such independent behaviour on the part of a woman would not have been possible in the Franco zone, there is no reason to believe that relations between the American and Dolores reached the level of intimacy condemned in Francoist descriptions of the atheistic and 'communist' sexual depravity of the Republic.

As for the Russians, Soviet pilots seem to have lived an extremely disciplined life in Spain, though Soviet commissars comment dourly on the conduct of some Russian advisers. Drunkenness was a frequent complaint. 'Liaisons with women' were frowned on.[17] Their memoirs, however, say little of their social life, which was probably circumscribed by their own and, at the most, Spanish communist circles. The journalist Virginia Cowles managed to find an interpreter to enable her to talk to three Russian aviators who had been shot down over Brunete and made prisoners. They said they had been attracted by the high pay offered. One said he had had only six months' flying experience. 'Both mentally and physically he seemed a man of inferior quality for an air-force pilot', wrote Cowles in a comment which confirms the view that the second cohort of Russian pilots was of markedly inferior quality to the first. Another Russian who was in hospital was the cynosure of all eyes, for here was a member of those fearsome 'Red Hordes' as they were known in Franco's Spain.[18]

What was learnt from the Spanish War?

Of the three foreign air forces which took part in the Spanish war, the experience was of greatest value to the Luftwaffe. It enabled the Germans to refine their Heinkel and Dornier bombers and their Messerschmitt fighters. The German discipline of close study, attention to detail and relentless search for perfection gave them indispensable lessons about the use in war of signals, logistics, repair workshops, bombs and anti-aircraft weapons (the Condor Legion had anti-aircraft as well as anti-tank sections and perfected the the use in both roles of the famous 88mm gun in Spain). An endless stream of reports went back to Germany. The Luftwaffe learnt the best ways to transport men, equipment and to make sure that spare parts were available. It studied the strategic bombing of ports, cities and factories and especially the technique of close forward support of advancing troops. The Germans learnt how vulnerable ground troops and their transport were to air attack and how important artillery was when combined with dominion over the air. Close coordination with tanks and infantry could shatter the enemy's defence. It was absence of air cover and of all-terrain armoured transport for troops that led to the disastrous Italian advance against Guadalajara in 1937, whereas the rapid advance of Franco's armies to the Mediterranean in March and April 1938 was achieved by a high level of German and Italian air–ground coordination. Precise strafing of the enemy's ground troops, even without the use of radio guidance until much later in the war, proved highly effective in the battles of 1937 and 1938, particularly in the 1937 campaign in the mountainous and difficult terrain of northern Spain as well as at Brunete in July 1937 and on the Ebro between July and November 1938. Much of the success was owed to efficient reconnaissance with Heinkel-45 and Heinkel-70 aircraft, which produced superb photos of, for instance, the 'Iron Belt' fortifications around Bilbao. The battlefield maps constructed from aerial photography allowed tactical carpet bombing according to detailed plans. Enemy troop movements could be detected, reported by radio and bombed within thirty minutes of the reconnaissance flight. Another of the advantages of techniques learned by the Luftwaffe in Spain was that air force commanders became intimately familiar with the front on the ground and functioned in close liaison with army headquarters. The Condor Legion assigned air force officers both to ground headquarters and to the front-line units themselves so that a senior airman was present at a focal point of the actual offensive. This helped develop and perfect techniques of air-to-ground communications, the identification of ground forces from the air and extremely close control of bombing and other attack from the air, and thus to minimize the risk of casualties from 'friendly fire'.[19] Above all, it was the veterans of the Condor Legion, Galland, von Richthofen, Jaenecke, Mölders (a general at the age of 29!), Harlinghausen and Plocher, all of whom later reached general rank, who organized the high degree of ground-to-air cooperation which reached full efficiency when total air–ground radio communication became available in the Second World War. The German Ju-87 dive-bomber was tried out experimentally in Spain, and von Richthofen, chief of staff and later commander of the Condor Legion, went on to command Ju-87 *Stuka* units in 1940.

In fighter techniques, in Spain the Luftwaffe developed its tactics of seeking height advantage, escaping attack by diving and regaining height and coming out of the sun to attack an enemy. It was Condor Legion pilot Werner Mölders who introduced the *Schwarm* or 'Finger-Four' technique which gave greater flexibility in combat than the traditional 'Vic' formation and was finally adopted by the British RAF in the Second World War. In 'Finger-Four', two machines (a *rotte*) flew at slightly different altitudes. One was the leader and the other the wingman. Another *rotte* flew slightly behind. The first wingman flew on the sun side of his leader, while the second watched the sky around the sun. The formation ensured that the leaders were guarded and collisions were more easily avoided than in the tighter RAF 'V' style'.[20] The *schwarm* increased lateral and vertical space between aircraft and, by giving less importance to tight formation flying, it allowed for more independence for the 125 German officers and 280 non-commissioned fighter pilots who flew in Spain. They achieved 314 confirmed and 70 probable victories, while sixty-one Republican aircraft were shot down by anti-aircraft fire.[21]

One German error may well have been not to have taken closer note of the relative lack of strategic value in bombing Spanish cities, even though the failure to destroy civilian morale had been evident. On 10 November 1932, British Prime Minister Baldwin had said in the House of Commons, 'The bomber will always get through.' But by the end of the war in Spain, even with the horrific experience of the three-day Italian bombing of Barcelona in March 1938 with the close to three thousand dead it caused, it was noticed that unescorted bombers were vulnerable to good defences and that properly organized air raid precautions could reduce civilian casualties. High-level bombing, in any case, was of doubtful accuracy. British observers noticed that even intensive bombing did not cause mass panic among civilians, and that deep shelters were effective in saving lives. Most observers concluded that Spanish morale remained high under bombing.[22] Bombing British cities in the Second World War, especially when air raid precautions were highly developed, never achieved the level of chaos and destruction that had been assumed. Indeed, it took massive and day and night bombing of German cities by huge fleets of hundreds of British and US aircraft to contribute to the German collapse in 1945. The German decision, to a considerable extent based on their experience in Spain, was to design the Luftwaffe to support the army rather than fight a strategic war on its own. Consequently, Germany did not develop a long-range strategic bomber.

Another German error was to assume that bombers such as the He-111 and the Dornier-17 were faster than enemy fighters and therefore could survive without fighter escorts. It was an attractive belief at the time, because fighters did not have sufficient range until the later Messerschmitts Bf-109 and Bf -110 were equipped with extra tanks and became able to carry out long-distance escort duties. But only when advanced bomber formations were designed and bombers became highly defended 'fortresses', could fighter escort be dispensed with.

As for the Soviets, regular reports were sent from Spain to Moscow, and these in turn were digested and translated into reports for the Commissar for Defence and in turn to the members of the Political Bureau of the Central Committee of the Communist Party of the Soviet Union.[23] Russian military advisers who had served with the Republic army in Spain lectured on their experiences. There was evidently no lack of published or otherwise-produced material. How much was absorbed into practical changes is, however, doubtful. As for lessons specifically from the war in the air, the conclusions were that for fighters speed was more important than manoeuvrability, that radio communication was vital, that actions on the ground required dominion of the air and that airfields should be well camouflaged. Other recommendations were that fast fighters were needed for bomber escort duties and that photography of the enemy's ground dispositions was necessary.[24] In other words, the recommendations were made, but whether they were incorporated into the flying techniques and manuals of the Soviet air force is questionable.

The USSR sent 648 aircraft altogether to Spain, including some of their finest fighters and the brilliantly conceived Tupolev SB-2 fast *Katiuska* bomber, which dominated the skies over the winter of 1936–7, together with 772 pilots. The rate of arrivals of Russian machines, however, declined from 1937 onwards, at the moment when Germany and Italy were increasing their air forces in Spain. In July 1937, one of the principal Soviet advisers sent an urgent message to Moscow:

> If the [Republican] army is to take the offensive and win, it will need sufficient fighters and bombers [. . .] At the least, the Republic needs 110–120 SBs and about 200–250 fighters, in addition to those it already has.[25]

Not until late December 1938, close to the end of the war, in response to Negrín's urgent message transmitted by Hidalgo de Cisneros, the Republican air force commander, did Stalin react to what he must or should have known was the parlous condition of the Spanish Republic's air force. It was too late, and about twice as many German and Italian aircraft served Franco's cause as Russian machines equipped the Republic.[26]

Furthermore, it was noticeable that the proportion of Soviet aircraft lost in accidents was intolerably high, indicating poor training. Between 25 October 1936, when the first Russian pilots arrived, and 1 July 1938, when most had left, 147 planes had been lost in forced landings in enemy territory and accidents, compared with only thirteen similar enemy losses.[27] Furthermore, whatever the alarm caused by the unexpected appearance of the *Katiuska*, it was in fact quite vulnerable and poorly protected once a highly skilled pilot such as Joaquín García Morato and the able Nationalist pilots he trained learned how to deal with it. In any case, Soviet mechanics reported that many of its parts were badly finished. As for fighters, the Soviet I-16 *Mosca* fighter, so advanced in 1936 that the Germans had to withdraw their He-51 from fighter duties, was becoming obsolescent by 1938–9, while the Messerschmitt Bf-109 was steadily developing ever faster and better-armed versions.

At the highest level, there does not appear to have been any agreement about the relationship between the Soviet senior commanders and the Spanish air staff. Indalecio Prieto, the Spanish Republic's Air Minister and later Minister of Defence, reported to his party that the Soviets behaved as if they were a law unto themselves. He claimed that Soviet advisers had insisted on bombing targets, which Prieto had rejected, and failed to bomb other cities which the minister had ordered.[28] The result was that the Soviet air force adviser, probably Lopatin, was withdrawn by Moscow and later purged.[29]

The uncertainty of Soviet military policy emerges clearly from the vicissitudes of some of the Russian pilots, who became Heroes of the Soviet Union and enjoyed successful careers after 1945, and many others who were either executed in the great 'Purges' of 1937–8 or paid with their lives for the disaster than befell the insufficiently developed Soviet air force in June 1941, among them Yakov Smushkevich, the first commander of Russian aircraft in Spain. In contrast, it was quite clear that the Condor Legion was in Spain to do what Franco wanted it to do and there was no question of political interference on the German part. In post-Communist Russia, the view is that there was lack of action and skill in the practical application of lessons learned in Spain, one of which was that the high quality of Soviet aircraft did not mean that no further development was needed. In contrast, German Condor Legion fighters steadily increased their speed and the calibre and range of their guns. Not until 1940 did the Soviet Yak and the MiG fighter prototypes appear, while the fast German Junkers 88 bomber was superior to the *Katiuska*. The USSR paid dearly for its delay at a time of rapid warplane development.[30]

Italy invested far more than it could afford, in aircraft as in other military resources, but it failed to create a staff like *Sonderstab* W which could coordinate the activities of land, sea and air forces. The irregular structure of command, in which Mussolini and his Foreign Minister Count Ciano rather than senior air force officers made decisions, did not encourage experimentation. Thus, there was little of the profound study engaged in by those German officers in Berlin who received the reports of the Condor Legion.[31] In particular, Italy confided overmuch in the ability of nimble fighter biplanes to shoot down bombers. Indeed, the very robust and manoeuvrable Fiat CR-32 *Chirri* fighter was the aircraft which maintained its value through the war, particularly since it was armed with a higher-calibre machine gun than the Russian fighters, so much so that Italy misjudged and assumed that it could continue to be useful in the Second World War, by when it was outdated by Spitfires and Hurricanes. These British fighters were armed with more and more powerful guns, as well as armoured protection for the pilots. The manoeuvrability of earlier biplanes such as the CR-32 became less important than the resistance of the fighter to the rapid and violent climbs and dives that were required.

The Italian air force leaders were convinced of the effectiveness of their bombing, although they recognized that in a civil war one side could not aim at the complete destruction of the enemy's cities.[32] In any case, in the Second World War Italy would have few chances to demonstrate its bombing power.

One of the principal questions about air war was whether the bomber would always get through, or whether it was highly vulnerable to the fighter. What sort of aircraft should a nation concentrate on manufacturing? The latest bombers which went through their paces in Spain could be tackled only by the newest, fastest and heaviest armed fighters. Though the *Katiuska* turned out to be vulnerable, the faster and better-finished Heinkel-111 bomber could be efficiently challenged only by the Soviet I-16, and no Italian Savoia-Marchetti S-79 bomber was shot down in Spain. So, despite the general view in Britain that not much could be learned from the Spanish Civil War, Air Chief Marshal Hugh Dowding of the Royal Air Force was undoubtedly right at the time to advance the cause of anti-aircraft defences and the mass building of the all-metal heavily armed and extremely fast Spitfire and Hurricane monoplane fighters, though probably it was only geography, radar and the efficient system of plotting which allowed RAF fighters the brief time that gave them the edge to get to a higher altitude than the German fighters.[33]

Who gained maximum advantage from the Spanish Civil War in the air? Of the non-Spanish air forces, it was Germany, whose fulminating victories of 1939 and 1940 were owed to the practical application in Spain of years of intense research into how warplanes should be used. It was Italy who gained the least advantage while, had the Soviet Union learned more than it did, it might have suffered less when Germany attacked it in June 1941.

Soviet intervention gained the USSR no political benefit either, because it failed to convince Britain and France to face up to the dictators. Nor did intervention in Spain serve Germany and Italy politically in the long term, because Franco did not help them to any meaningful extent in the Second World War.

It was Franco and the Spain he created which most profited from German and Italian aid, of which aircraft and air crew were perhaps the most valuable part. Without German and Italian aid, the Nationalists would not have won the war and instituted a regime which lasted until 1975..

Afterthought

Reflections on Spanish 'memory' eight decades later

Since the death of Franco in November 1975 and the transition to democracy, Spain has largely, though not completely, rejected its Francoist past. Nevertheless, the emotions and resentments aroused by the war of 1936–9 are still raw. Two specific and major matters of tension remain. One is the location, disinterment and reburial of the remains of people executed by the insurgents with either no trial or after mere simulacra of legal process, and the other is the responsibility for the destruction of Guernica on 26 April 1937.

Once the lie that the Basques themselves had destroyed Guernica was no longer sustainable, the responsibility for the outrage was thrown more and more on the Condor Legion, until, following Franco's death, historians began to investigate the responsibilities of the Nationalist command itself. Did General Mola, commander of the Insurgent Army of the North, and his chief of staff, Colonel Vigón, request the bombing in the particular savage form that it took, or did they simply turn a blind eye to the consequences of the type of attack that the Condor Legion had been developing in Spain? So far, no reliable evidence has emerged to answer these questions.

In post-1945 Germany, the Nazi associations of the Condor Legion were frowned on and few ex-Condor Legion men joined that force's old comrades' association, even if some of the younger officers who survived the Second World War achieved high rank in the post-1955 Luftwaffe. In 1997, on the sixtieth anniversary of the bombing, the German president, Roman Herzog, sent a message to be read in Guernica by the ambassador to Spain, in which he admitted the culpability of German airmen. In more recent years, the role and responsibility of the entire German military machine – the *Wehrmacht* – for Nazi behaviour, including the question of the 'legitimacy' of the German intervention in Spain and the matter of the bombing ethic of the Luftwaffe, which required its pilots to ignore collateral harm to civilians, as exemplified by the destruction of Guernica, are under discussion. For the moment, one might conclude that Fascist and Nazi regimes, almost by definition, did not act in a way which could be considered legitimate, and that their behaviour was reflected in the actions of their coercive forces, both police and military.

Notes

Preface

1 This book uses the term 'Republican' to refer to the forces defending the Spanish Republic. 'Insurgents', 'Nationalists' or 'Francoists' refer to Franco's forces.
2 Raymond L. Proctor, *Hitler's Luftwaffe in the Spanish Civil War* (Westport, CT: Greenwood Press, 1983), p. 248. Also Stephanie Schüler-Springorum, *La guerra como aventura: la Legión Cóndor en la guerra civil española 1936–1939* (Madrid, 2014), translated from *Krieg und Fliegen. Die Legion Condor im Spanischen Bürgerkrieg* (Paderborn, 2010), p. 313.
3 Quoted in Gerald Howson, *Arms for Spain: The Untold Story of the Spanish Civil War* (London: John Murray,1998), p. 15.
4 An account of how the Great Powers reacted to the Spanish war can be found in Michael Alpert, *A New International History of the Spanish Civil War*, 2nd ed. (Basingstoke: Palgrave, 2004).
5 I thank Sebastian Balfour for clarifying this matter.
6 For a general view of the beginning of the civil war, see Hugh Thomas, *The Spanish Civil War* (London: Penguin, 2012); Paul Preston, *A Concise History of the Spanish Civil War* (London: Fontana, 1996); and Anthony Beevor, *The Spanish Civil War* (London: Penguin, 1982).
7 *Air War over Spain* (translated by Margaret A. Kelley) (London: Ian Allan, 1974).

Chapter 1

1 J. A. Ansaldo, in J. L. Vila San Juan, *Enigmas de la guerra civil española* (Barcelona, 1974), p. 17.
2 On Franco the major work available is Paul Preston, *Franco* (London: HarperCollins, 1993).
3 Anonymous (Luis Bolín), Marquis del Moral and Douglas Jerrold, *The Spanish Republic: A Survey of Two Years of Progress* (London: Eyre and Spottiswoode, 1933).
4 Douglas Jerrold, *Georgian Adventure* (London: Collins, 1937), pp. 371–2.
5 Preston, *Franco*, pp. 138–9.
6 Gerald Howson, *Aircraft of the Spanish Civil War, 1936–1939* (London: Putnam, 1990), p. 105.
7 H. Thomas, *The Spanish Civil War* (London: Penguin, 2012), pp. 224–5.

8 For a series of investigations into the illegitimacy of the military insurrection, see Angel Viñas et al., *Los mitos del 18 de julio* (Barcelona, 2013).
9 Chapter on Núñez del Prado in J. García Fernández (ed.), *Veinticinco (25) Militares de la República* (Madrid, 2011), pp. 727–52, specifically pp. 749–52.
10 Details in Howson, *Aircraft of the Spanish Civil War*, p. 38; and Preston, *Franco*, pp. 278–9.
11 For the Spanish navy in the civil war, see Michael Alpert, *La guerra civil española en el mar* (Barcelona, 2007).
12 Carlos Lázaro Avila 'Ignacio Hidalgo de Cisneros', in *25 militares de la República* (Madrid, 2011), pp. 505–42, specifically p. 522.
13 Quoted in Michael Alpert, *la reforma militar de Azaña* (Granada, 2nd ed., 2008), p. 207.
14 The Spanish air force has squadrons (*escuadrillas*), wings (*grupos*) and groups (*escuadras*) in ascending order of size.
15 These and other technical details, together with numbers of aircraft in service, are taken from Howson, *Aircraft of the Spanish Civil War* (listed alphabetically), contrasted with Jesús Salas Larrazábal, *La guerra civil desde el aire*, and Andrés García Lacalle, *Mitos y verdades: la aviación de caza en la guerra española* (Mexico City, 1973), pp. 18–22.
16 Details about the role of Major Ignacio Hidalgo de Cisneros between 1935 and the outbreak of the civil war are taken from his *Cambio de Rumbo*, 2 vols. (Bucharest, 1964), ii, pp. 117–41.
17 For essays on Camacho and Cascón, see *25 militares de la República*, pp. 195–211 and 261–5.
18 Luis Romero, *Tres días* de *julio* (Esplugues de Llobregat, 1967), p. 398.
19 M. Koltsov, *Diario de la guerra de* España (1st publ., Moscow, 1957; Paris, 1963), p. 5.
20 Full details of the vicissitudes of army officers can be found in Michael Alpert, *The Republican Army in the Spanish Civil War 1936–1939* (Cambridge: Cambridge University Press, 2013), pp. 86–91, 319–21.
21 Quoted from García Lacalle, *Mitos y verdades*, pp. 71–82. For more details on Tablada, see the very tendentious J. Arrarás, *Historia de la Cruzada Española* (Madrid, 1941), Vol. 3, Book 11, pp. 166–7, 211. See also Gerald Howson, *Arms for Spain: The Unknown Story of the Spanish Civil War* (London: John Murray, 1998), pp. 10–11.
22 See the protest by General Kindelán, on 26 November 1936, in the latter's *Mis cuadernos de guerra* (Barcelona, 1982), p. 47.
23 Jesús Salas Larrazábal, *La guerra civil desde el aire*, pp. 62–3; and Howson, *Aircraft of the Spanish Civil War*, p. 20.
24 Hidalgo de Cisneros, p. 153. Salas, *La guerra civil desde el aire*, p. 59, gives 200 out of 500.
25 García Lacalle, *Mitos y verdades*, p. 15.
26 C. Engel Masoliver, *El cuerpo de oficiales en la guerra de España* (Valladolid, 2008).
27 *Gaceta de la República*, No. 50, 19 February 1937.
28 García Lacalle, *Mitos y verdades*, pp. 121–2.
29 J. Salas Larrazábal, *La guerra civil desde el aire*, p. 59, gives 200 out of 500.
30 J. Salas Larrazábal, Appendix 13 and cp. *Anuario Militar de España* (annual military list) for 1936.
31 García Lacalle, *Mitos y verdades*, p. 15.
32 Ibid., p. 16.

33 Ibid.
34 Ibid., pp. 14–15.
35 On the navy, see Michael Alpert, *La guerra civil española en el mar*, p. 107, and, by the same author, 'The Clash of Spanish Armies: Contrasting Ways of War in Spain 1936–1939', *War in History* 6, 3 (1999), pp. 331–51.
36 J. Salas Larrazábal, *La guerra civil desde el aire*, p. 73; and García Lacalle, *Mitos y verdades*, p. 15.
37 García Lacalle, *Mitos y verdades*, p. 18.
38 Ibid., p. 23.
39 All loyal officers were to receive one promotion.
40 José Larios, *Combat over Spain: Memoirs of a Nationalist Pilot 1936–1939* (London: Neville Spearman, 1968), p. 77.

Chapter 2

1 Michael Alpert, *La guerra civil española en el mar* (Barcelona: Crítica, 2007), pp. 43–55, gives an account of mutinies in the Spanish fleet.
2 The major work on this aspect is Angel Viñas, *Franco, Hítler y el estallido de la guerra civil española: antecedentes y consecuencias* (Madrid: Alianza Editorial, 2001).
3 *Documents on German Foreign Policy 1918–1945*, Vol. 3, Series D (London: His Majesty's Stationery Office, 1951), Document 2 (this source will be referred to as *GD*).
4 M. Alpert, *A New International History of the Spanish Civil War* (Basingstoke: Palgrave, 2004), pp. 27–41, gives details of the meeting with Hitler.
5 Stefanie Schüler-Springorum, *La guerra como aventura: la Legión Cóndor en la guerra civil española 1936–1939* (Madrid: Alianza Editorial,2014), p. 54, note 13.
6 Gerald Howson, *Aircraft of the Spanish Civil War, 1936–1939* (London, Putnam, 1990), p. 207; and Andrés García Lacalle, *Mitos y verdades: la aviación de caza en la guerra española* (Mexico City: Oasis, 1973), p. 118.
7 Quoted in H. M. Mason, *The Rise of the Luftwaffe* (London: Cassell, 1975), p. 125.
8 For a detailed study of the rebirth of German air power after 1918, see James S. Corum, *The Luftwaffe: Creating the Operational Air War, 1918–1940* (Lawrence: University Press of Kansas, 1997).
9 Mason, *Rise of the Luftwaffe*, p. 164.
10 Corum, *The Luftwaffe*, p. 59.
11 Stephen Budiansky, *Air Power* (New York: Viking, 2003), pp. 203–5. See also the excellent summary in *The Rise and Fall of the German Air Force, 1933–45* (Air Ministry, London, 1948, and republished by National Archives 2001 from the original in AIR 41/10. See also, Corum (1997).
12 Corum, *The Luftwaffe*, p. 190.
13 Schüler-Springorum, *La guerra como Aventura*, p. 51.
14 E. R. Hooton, *Phoenix Triumphant: The Rise and Fall of the Luftwaffe* (London: Brockhampton Press, 1999), p. 122.
15 *Hitler's Table Talk 1941–44; His Private Conversations*, ed. H. R. Trevor-Roper (London: Weidenfeld and Nicolson, 1953), p. 687.
16 National Archives, Kew, HW series No. 065678.
17 National Archives, Kew, Foreign Office General Correspondence W8263/62/41.
18 Alpert, *New International History of the Spanish Civil War*, p. 32.

19 Arthur Koestler, *Spanish Testament* (London: Gollancz, 1937), p. 37.
20 H. Ries and K. Ring, *The Legion Condor* (trans. D. Johnston) (West Chester, PA: Schiffer, 1992), p. 16.
21 Koestler's account of his time in prison is in his *Dialogue with Death*, the only part of *Spanish Testament* to be republished (London: Collins and Hamish Hamilton, 1954).
22 Schüler-Springorum, *La guerra como aventura*, p. 95.
23 A. Galland, *The First and the Last* (London: Methuen, 1955), p. 23.
24 Schüler-Springorum, *La guerra como aventura*, pp. 100–3, based on diaries.
25 Francisco Sánchez Ruano, *Islam y la guerra civil española* (Madrid, 2004), p. 138.
26 R. Proctor, *Hitler's Luftwaffe in the Spanish Civil War* (New York: Praeger, 1983), pp. 35–6.
27 *GD*, nos. 92 and 94.
28 Håkan's Aviation Page (surfcity.kund.dalnet.se/scw), 15 September 1936.
29 Alpert, *New International History*, p. 83.
30 Soviet aircraft will be discussed below. Technical details are drawn from Howson, *Aircraft of the Spanish Civil War*, pp. 276–7.
31 Proctor, *Hitler's Luftwaffe in the Spanish Civil War*, p. 55.
32 This account of the formation of the Condor Legion is drawn from Proctor, chapter 5 and 'The Luftwaffe in the Spanish Civil War 1936–1939', in Corum, *The Luftwaffe*, pp. 182–223.
33 Proctor, *Hitler's Luftwaffe in the Spanish Civil War*, p. 57.
34 Corum, *The Luftwaffe*, p. 191.
35 *GD*, No. 113.
36 Alpert, *New International History of the Spanish Civil War*, pp. 95–6.
37 Proctor, *Hitler's Luftwaffe in the Spanish Civil War*, p. 59, note 8.
38 Ibid., p. 60.
39 Ibid., p. 69.

Chapter 3

1 See John F. Coverdale, *Italian Intervention in the Spanish Civil War* (Princeton: Princeton University Press, 1975, pp. 50–4.
2 Angel Viñas,'La connivencia fascista con la sublevación y otros éxitos de la trama civil' ('Fascist connivance with the uprising and other successes of the civilian network'), in A. Viñas (ed.), *Los mitos del 18 de julio* (Barcelona, 2013), pp. 79–181.
3 For a more detailed account of the process of decision in Rome, see M. Alpert, *A New International History of the Spanish Civil War* (Basingstoke: Palgrave, 2004), pp. 35–9.
4 See John Gooch, *Mussolini and His Generals: The Armed Forces and Fascist Foreign Policy, 1922–1940* (New York: Cambridge University Press, 2007), pp. 372–3.
5 Alpert, *A New International History*, p. 37.
6 Edoardo Grassia, '"Aviazione Legionaria": il comando strategico-politico e tecnico-militare delle forzee aeree italiane impiegate nel conflitto civile spagnolo', *Diacronie:Studi di Storia Contemporanea* 7, 3 (2011), pp. 1–24 (specifically pp. 5, 10 and 17). I thank the author for sending me this publication.
7 The names of the members of each crew are given by Guido Matteoli, *L'Aviazione legionaria in Spagna* (Rome,1940), p. 15.

8 Details are taken from Ferdinando Pedriali, *Guerra di Spagna e Aviazione italiana* (Rome,1992), chapter 3, based on Colonel Bonomi's own reports to General Valle.
9 Lucia Ceva, 'Conseguenze dell'intervento italo-fascista', in Sacerdoti Mariani, A. Colombo and A. Pasinato (eds) *La guerra civile spagnola tra politica e letteratura* (Florence, 1995), 1995, pp. 215–39, specifically p. 218. Also Brian Sullivan, 'Fascist Italy's Military Involvement in the Spanish Civil War' *Journal of Military History* 59, 4 (1995), pp. 697–727, specifically p. 718.
10 *Documents Diplomatiques Français*, 1932–9, Series 2 (1936–9), Vol. III, No. 46 (Paris, 2003), dispatch to Paris from the French Resident-General in Rabat.
11 For Blum's reaction see Alpert, *A New International History*, p. 41.
12 This complicated story can be followed in Alpert, *A New International History*, chapter 4.
13 Quoted verbatim by Pedriali, *Guerra di Spagna e Aviazione Italiana*, p. 41.
14 For German ships, see M. Alpert, *La guerra civil española en el mar* (Barcelona, 2007), p. 95.
15 Figures for the aircraft that were available to protect the convoy vary. My details come from J. M. Martínez Bande, *La campaña de Andalucía* (Madrid, 1969), p. 36, note 31 and Document 2, which is the relevant order to the aircraft issued at 11 pm on 4 August.
16 Ibid. Document 3 and Pedriali, *Guerra di Spagna e Aviazione italiana*, p. 45.
17 Alpert, *La guerra civil española en el mar*, pp. 98–9.
18 Ibid., p. 100.
19 Martínez Bande, *La campaña de Andalucía*, p. 43, note 41.
20 Ibid., p. 43.
21 For the brutality of the insurgent columns see P. Preston, *The Spanish Holocaust* (London: Harper Press, 2013), chapter 9, and for an analysis of its causes see Sebastian Balfour, *Deadly Embrace: Morocco and the Road to the Spanish Civil War* (Oxford: Oxford University Press, 2002), pp. 280–1.
22 Preston, *Spanish Holocaust*, pp. 318–19.
23 Pedriali, *Guerra di Spagna e Aviazione italiana*, p. 52.
24 *Ciano's Diary 1939–1943* (London: Heinemann, 1947), p. 33.
25 J. Larios, *Combat over Spain; Memoirs of a Nationalist Fighter Pilot 1936–1939* (London: Neville Spearman, 1968), p. 149.
26 Pedriali, Guerra di Spagna e Aviazione italiana, p. 53 and note 8 containing reports from Colonel Bonomi.
27 Ibid., p. 56.
28 Gerald Howson, *Aircraft of the Spanish Civil War, 1936–1939* (London: Putnam, 1990), p. 134.
29 Pedriali, *Guerra di Spagna e Aviazione italiana*, p. 54, note 10. No date is given.
30 Ibid., p. 56.
31 Ibid., p. 58. Normally, international law requires internment of a belligerent who reaches a neutral country, but it could be argued that this does not apply in a civil war.
32 See 'Hispanicus', *Foreign Intervention in Spain* (London: United Editorial, undated), Vol. 1, pp. 212–16; and Judith Keene, *Fighting for Franco* (London: Leicester University Press, 2001), pp. 95–9.
33 Alpert, *La guerra civil española en el mar*, p. 107.
34 Gerald Brenan, *Personal Record 1920–1972* (New York: Knopf, 1975), 308–9.
35 Sir P. Chalmers-Mitchell, *My House in Málaga* (London: Faber and Faber, 1938), pp. 182–3.

36 Luis Miguel Cerdera, *Málaga, base naval accidental* (Seville, 2015), pp. 154–60.
37 J. Salas Larrazábal, *La Guerra de España desde el aire*, p. 106
38 Joaquín García Morato, *Guerra en el Aire* (Madrid, 1940), p. 92 This ace fighter pilot was killed in a flying accident on 4 April 1939 just after the war ended.
39 Details of March's operations through Kleinwort-Benson come from Jehanne Wake, *Kleinwort-Benson: The History of Two Families in Banking* (Oxford: Oxford University Press, 1997), pp. 252–3.
40 Document No. 5 of the official Francoist report, cited by J. M. Martínez Bande, *La invasión de Aragón y el desembarco en Mallorca* (Madrid, 1970), which does not mention Juan March, though most other sources do (See Coverdale, *Italian Intervention in the Spanish Civil War*, p. 132). Pedriali, *Guerra di Spagna e Aviazione Italiana*, p. 69, note 14, claims that the total cost was far greater than the amounts subscribed by March and on the island itself. See also Mercedes Cabrera, *Juan March* (Madrid, 2011), pp. 294–5.
41 Cabrera, *Juan March*, p. 298.
42 Pedriali, *Guerra di Spagna e Aviazione Italiana*, p. 78.
43 For a detailed consideration of the international repercussions of the Italian presence on Majorca, see Coverdale, *Italian Intervention in the Spanish Civil War*, chapter 5.
44 Pedriali, *Guerra di Spagna e Aviazione Italiana*, p. 92.
45 For the increase in Italian aid to Franco and British reaction, see Alpert, *A New International History*, pp. 90–4.

Chapter 4

1 Blum's account of his dealings with Republican Spain is taken from his post-Second World War report to the Assemblée Nationale (*Rapport fait au nom de la commission chargée d'enquêter sur les évènements survenus en France de 1933 à 1945*, Paris, 1951). The French text of Giral's telegram is quoted in a number of very slightly differing versions, which suggests that it has never been found. One version reads, *Sommes surpris par dangereux coup de main militaire.Vous demandons de vous entendre immédiatement avec nous pour fournitures d'armes, avions. Fraternellement, Giral*. See M. Alpert, *A New International History of the Spanish Civil War* (Basingstoke: Palgrave, 2004), Chapter 2; and J.-F. Berdah, *La Democratie assassinée: La République espagnole et les grandes puissances, 1931–1939* (Paris, 2000), p. 216.
2 The quantities of war material vary according to the source. My figures are quoted from official French documents as cited by Gerald Howson, *Arms for Spain: The Untold Story of the Spanish Civil War* (London: John Murray, 1998), p. 23.
3 Angel Viñas, *El Oro de Moscú* (Barcelona, 1979), pp. 36–7.
4 Given the extremely detailed research of Gerald Howson (*Aircraft of the Spanish Civil War, 1936–1939* [London: Putnam, 1990], p. 254), who argues that the larger numbers often mentioned are unreliable, I would opt for his figure of eighteen Potez aircraft.
5 Ibid., Appendix I. The French document proves that most of the aircraft were flown to Barcelona on 8 August.
6 Ibid., p. 112. Again, Howson's research does not support the considerably larger number of Dewoitine fighters mentioned by Francoist historians and those who have copied their figures.

7 Ibid., p. 48, shows that Spain was charged about double what Lithuania had been going to pay.
8 Ibid., p. 55.
9 I am again reliant on Howson, *Aircraft of the Spanish Civil War*, p. 254, because of the thoroughness of his convincing research.
10 Andrés García Lacalle, *Mitos y verdades: la aviación de caza en la guerra española* (Mexico City, 1973), p. 19; and Howson, *Aircraft of the Spanish Civil War*, p. 256.
11 Ibid., p. 59.
12 Robert S. Thornberry, *André Malraux et l'Espagne* (Geneva, 1977), pp. 19–21.
13 Ibid., pp. 24–8.
14 See the well-documented thesis of Richard A. Cruz, *André Malraux: The Anticolonial and Antifascist Years* (Denton: University of Texas digital library, 1996), chapter 8, pp. 208–31. Accessed on 4 July 2016 at http/:digital.library.edu.ark/67531/metadc 277996.
15 Thornberry, *André Malraux et l'Espagne*, Appendix I, provides lists of the members of the *Escadrille Espagne* for August–November 1936 and November 1936–February 1937.
16 See photograph of Jean Dary's contract in Jean Gisclon, *Des aviones et des hommes* (Paris, 1969), opposite p. 128. Some reports quote 50,000 pesetas, but that sum at the time represented the enormous amount of £2,000, sterling while 50,000 francs, even when on 26 September the franc was devalued to 105 to the pound sterling, was worth the much smaller, though still generous for 1936, sum of £476.
17 Gisclon, *Des aviones et des hommes*, p. 33.
18 Howson, *Arms for Spain*, Appendix I, prints the lists.
19 Gisclon, *Des aviones et des hommes*, p. 28.
20 Sometimes spelt 'Darry'. Most of the personal information on the pilots is taken from Gisclon, *Des aviones et des hommes*, pp. 42–50.
21 García Lacalle, *Mitos y verdades*, p. 150.
22 Ibid.
23 M. Koltsov, *Diario de la guerra de España* (Paris, 1963), p. 11.
24 Cruz thesis, p. 215. See also R. Skoutelsky, *Novedad en el frente* (Madrid, 2006), pp. 45–8.
25 'J.D' in 'La guerre aérienne en Espagne', *Revue de l'Armée de l'air'* (No. 96, July 1938, pp. 808–24), specifically p. 816.
26 Gisclon, *Des aviones et des hommes*, p. 35.
27 Ibid., p. 36.
28 Cruz thesis, p. 216.
29 J. Salas Larrazábal, *La guerra de Espoaña desde el aire*, p. 94.
30 Cruz thesis, p. 217.
31 Different sources give varying dates for this action, but Thornberry, *André Malraux et l'Espagne*, pp. 43–4, on the basis of military reports, prefers 16 August.
32 Ibid., p. 218. Howson, *Aircraft of the Spanish Civil War*, p. 255, gives the date as 16 August, while J.Salas, *La Guerra de España* (p. 95) gives no date and says only that there was an efficient bombing raid which obliged the insurgents to send more fighters to the area.
33 Howson, *Aircraft of the Spanish Civil War*, p. 254.
34 Koltsov, pp. 120–2.
35 According to a report by Dary to Pierre Cot referring to early October but dated 21 November 1936, there were three Dewoitine 372s, two Loire 46 fighters (which had

arrived in early September) and one sole Hawker Fury left. Dary had given a copy of his report to the British Air Attaché. It can be found in the National Archives (Kew) in FO371/21284/W2999.

36 Gisclon, *Des aviones et des hommes*, p. 226.
37 Howson, *Arms for Spain*, pp. 102–3.
38 Quoted in Thornberry, *André Malraux et l'Espagne*, p. 45.
39 Ignacio Hidalgo de Cisneros, *Cambio de Rumbo* (Bucharest, 1964), ii, pp.173–4.
40 Amanda Vaill, *Hotel Florida: Truth, Love, and Death in the Spanish Civil* War (Bloomsbury, 2014), vividly portrays the atmosphere in wartime Madrid.
41 Hidalgo de Cisneros, *Cambio de Rumbo*, ii, p. 174.
42 Thornberry, *André Malraux et l'Espagne*, p. 41.
43 Ibid., p. 50.
44 For the commissars in the forces of the Republic see Michael Alpert, *The Republican Army in the Spanish Civil War, 1936–1939* (Cambridge: Cambridge University Press, 2013), Chapter 8, pp. 174–201.
45 Thornberry, *André Malraux et l'Espagne*, p. 52.
46 Paul Nothomb, *La Rançon* (Paris, 2001), pp. 23–5.
47 For details of attempts to buy British aircraft, see Howson, *Arms for Spain*, pp. 90–7.
48 Much detail on these men is provided by Brian Bridgeman, *The Flyers* (Swindon: Brian Bridgeman, 1989), particularly chapter 2.
49 Howson, *Arms for Spain*, pp. 61–5.
50 Bridgeman, *The Flyers*, p. 91.
51 Ibid., pp. 44–6.
52 Oloff de Wet, *Cardboard Crucifix* (Blackwood, 1938), p. 12.
53 Bridgeman, *The Flyers*, p. 76.
54 Quoted Bridgeman, *The Flyers*, pp. 71–2, from a magazine article by de Wet.
55 Bridgeman, *The Flyers*, pp. 96–7.
56 Ibid., p. 121.
57 Ibid., pp. 122–3.
58 García Lacalle, *Mitos y verdades*, pp. 141–4.
59 Bridgeman, *The Flyers*, Chapter 6.
60 Martín-Lunas does not appear in the 1936 Spanish Military Yearbook, which suggests that he may have taken early retirement before the war.
61 García Lacalle, *Mitos y verdades*, p. 134.
62 Paul Whelan, *Soviet Airmen in the Spanish Civil War* (Atglen, PA: Schiffer, 2014), who publishes the Russian names in alphabetical order.
63 Published in Moscow, no date, pp. 367ff.
64 Sterling Seagrave, *Soldiers of Fortune* (Alexandria, VA: Time-Life, 1981), p. 4.
65 For Cascón, see *25 Militares de la República* (Madrid, 2011), pp. 263–93.

Chapter 5

1 R. Radosh, Mary Habeck and G. Sevostianov (eds.), *Spain Betrayed: The Soviet Union in the Spanish Civil War* (New Haven, CT: Yale University Press, 2001), Document 10.
2 Yuri Rybalkin, *Stalin y España: la ayuda militar soviética a la República* (Madrid, 2007), pp. 50–1.

3 Radosh et al., *Spain Betrayed*, Document 9.
4 For a detailed account of Soviet policy towards Spain in July–October 1936, see M. Alpert, *A New International History of the Spanish Civil War* (Basingstoke: Palgrave Macmillan), pp. 48–52.
5 *Documents on British Foreign Policy 1919–1939, 2nd Series, Vol. 17* (London: HMSO, 1979), Document No. 32.
6 Rybalkin, *Stalin y España*, pp. 60–1.
7 See R. W. Davies et al. (eds), *The Stalin–Kaganovich Correspondence, 1931–1936* (New Haven, CT: Yale University Press, 2003), Document 34.
8 Rybalkin, *Stalin y España*, p. 52.
9 Cited by D. Kowalsky, *La Unión Soviética y la Guerra Civil Española: una revisión crítica* (Barcelona, 2004), p. 202.
10 Alexander Boyd, *The Soviet Air Force since 1918* (London: MacDonald and Jane's, 1977), p. 75.
11 All details of Soviet shipments, arrivals and numbers of aircraft are taken from Gerald Howson's *Arms for Spain*, Appendix III, which is in turn taken from the Russian State Military Archives (RGVA). These data are contrasted with those of the most recent Spanish study: E. Abellán Agius, *Los cazas soviéticos en la guerra aérea de España, 1936–1939* (Madrid,1999), pp. 10–17. For additional information on the Soviet organization of military aid to Spain, see Rybalkin, *Stalin y España*, especially chapter 2. Information on Russian aid to the Spanish Republic in army and naval matters can also be found in Michael Alpert's *The Republican Army in the Spanish Civil War*, in the same author's *La guerra civil española en el mar* and in Kowalsky, *La Unión Soviética y la Guerra Civil Española*.
12 The dates and numbers are taken from E. Abellán Agius, *Los cazas soviéticos en la guerra aérea de España*, together with a letters from the late Gerald Howson to Paul Whelan of 12 April 2003 and to Frank Schauff of 25 April 2003, copies of which I was given by Howson.
13 Gerald Howson, *Aircraft of the Spanish Civil War, 1936–1939* (London: Putnam, 1990), p. 305. This source underestimates numbers of Italian aircraft by close on one hundred (see Pedriali, p. 390).
14 Howson, letter dated 12 April 2003 to Paul Whelan, author of *Soviet Airmen and the Spanish Civil War* (Atglen, PA: Schiffer, 2014).
15 For a discussion of the Soviet withdrawal from Spain as a lost cause, see Alpert, *A New International History*, pp. 184–5.
16 Ranks are quoted as in Paul Whelan's alphabetical list of *Soviet Airmen in the Spanish Civil War*.
17 Andrés García Lacalle, *Mitos y verdades: la aviación de caza en la guerra española* (Mexico City, 1973), p. 175.
18 García Lacalle, *Mitos y verdades*, p. 249.
19 Ignacio Hidalgo de Cisneros, *Cambio de Rumbo* (Bucharest, 1964), p. 187, elaborates on the murder of the pilot. García Lacalle, however, does not mention the incident. The Soviet pilot Prokofiev in *Bajo la bandera de la República española*, pp. 385–6, describes the event but says that Ghibelli was shot down south of Madrid. Whelan suggests that the pilot whose body was returned may have been the Russian Vladimir Bocharov.
20 Håkan's Aviation page – Air War in the Spanish Civil War 1936, October 1936.
21 For the later vicissitudes of Soviet airmen, see Whelan, *Soviet Airmen and the Spanish Civil War*, under names.
22 Whelan (under Kolesnikov).

23 Whelan (under Pumpur).
24 *Bajo la bandera*, p. 378.
25 Håkan's Aviation page, 23 November 1936.
26 *Bajo la bandera*, p. 380
27 Howson, *Aircraft of the Spanish Civil War*, p. 277, says that only four *Katiuskas* flew that mission. Prokofiev's account mentions six.
28 Joaquín García Morato, *Guerra en el aire* (Madrid, 1940), pp. 49–50. For the view from the Republican side, see García Lacalle, *Mitos y verdades*, pp. 290–1.
29 Håkan, November 1936.
30 García Lacalle, *Mitos y verdades*, p. 293.
31 Ibid., pp. 232–3.
32 Håkan, October 1936.
33 Ibid., 13 November 1936.
34 For the vicissitudes of Tupikov's reconnaissance flight, see Prokofiev in *Bajo la bandera*, pp. 381–4. For Tupikov, see Whelan under name.
35 Frank Tinker, *Some Still Live: Experiences of a Fighting-Plane Pilot in the Spanish Civil War* (London: Lovat Dickson, 1938), p.136.
36 See biography of Tinker by Richard K. Smith and R. Cargill Hall, *Five Down, No Glory* (Annapolis, MD: Naval Institute Press, 2011).
37 Ibid., p. 161.
38 This episode is in Tinker, *Some Still Live*, pp. 34–5.
39 Ibid., p. 36.
40 Ibid., pp. 85–6.
41 See Amanda Vaill, *Hotel Florida: Truth, Love and Death in the Spanish Civil War* (London: Bloomsbury, 2014). On Hemingway's earnings, see Herbert Mitgang, 'Hemingway on Spain: Unedited Reportage' *New York Times*, 30 August 1988.
42 Coverdale, Appendix C. The exact total was 47,176. On Guadalajara, see John F. Coverdale, *Italian Intervention in the Spanish Civil War* (Princeton, NJ: Princeton University Press, 1975), pp. 212–60.
43 See García Lacalle, *Mitos y verdades*, p. 239.
44 Ibid., p. 241.
45 Tinker, *Some Still Live*, pp. 124–5.
46 Ibid., pp. 131–2.

Chapter 6

1 Håkan's Aviation Page (Surfcity.kund. dalnet.se/cr32.htm), November 1936.
2 E. Hooton, *Phoenix Triumphant: The Rise and Fall of the Luftwaffe* (London: Brockhampton Press, 1999), p. 109.
3 R. Proctor, *Hitler's Luftwaffe in the Spanish Civil War* (New York: Praeger, 1983), p. 98. For other references see D. Baker, *Adolf Galland: The Authorised Biography* (London: Windrow and Greene, 1996), and Galland's own account *The First and the Last*: *The German fighter force in World War II* (London: Methuen, 1955).
4 Håkan, November 1936.
5 Ibid.
6 Translated from the Spanish translation of the original German and archived in the Archivo General Militar at Avila (hereafter AGMAV), 2218.2.9.

7 Håkan, November 1936.
8 Hooton, *Phoenix Triumphant*, p. 126.
9 Håkan, December 1936; italics added.
10 Håkan, January 1937.
11 Proctor, *Hitler's Luftwaffe in the Spanish Civil War*, p. 89, quoting the German pilot Douglas Pitcairn *verbatim*.
12 Ibid.
13 Ibid., pp. 83–5.
14 Ibid., p. 90.
15 Håkan, January 1937.
16 Often referred to simply as the Bf-109, after the original *Bayerische Flügzeugwerke* factory.
17 Proctor, *Hitler's Luftwaffe in the Spanish Civil War*, p. 90.
18 For the pre-Spanish Civil War history and the technical specifications of these bombers, see G. Howson, *Aircraft of the Spanish Civil War, 1936–1939* (London: Putnam, 1990).
19 See R. C. Smith and R. K. Hall, *Five Down, No Glory. Frank G. Tinker, Mercenary Ace in the Spanish Civil War* (Annapolis, MD: Naval Institute Press, 2011), p. 139. Tinker, unfamiliar with the new German aircraft, thought they were the new Junkers (p. 112).
20 See J. M. Martínez Bande, *Vizcaya* (Madrid, 1971), p. 39.
21 Proctor, *Hitler's Luftwaffe in the Spanish Civil War*, p. 118.
22 On the sinking of the *España* see M. Alpert, *La Guerra Civil Española en el mar* (Barcelona, 2007), pp. 278–9.
23 Ignacio Hidalgo de Cisneros, *Cambio de Rumbo* (Bucharest, 1964), ii, p. 225.
24 Howson, *Aircraft of the Spanish Civil War*, p. 217.
25 G. Howson, *Arms for Spain: The Untold Story of the Spanish Civil War* (London: John Murray, 1998), pp. 212–13.
26 For a list of the 'Circo Krone' aircraft, see E. Abellán, *Los cazas soviéticos en la guerra aérea de España* (Madrid, 1999), Appendix VIII.
27 Stefanie Schüler-Springorum, *La guerra como aventura: la Legión Cóndor en la guerra civil española 1936–1939* (Madrid, 2014), p. 68.
28 Proctor, *Hitler's Luftwaffe in the Spanish Civil War*, 124–5.
29 Ibid., p. 125.
30 Ferdinando Pedriali, *Guerra di Spagna e aviazione italiana* (Rome, 1992), p. 217.
31 J. M. Solé i Sabaté and Joan Villarroya, *España en llamas: la guerra civil desde el aire* (Madrid, 2003), pp. 79–80.
32 Quoted by George Lowther Steer in his *The Tree of Gernika: A Field Study of Modern war* (London: Hodder and Stoughton, 1938), p. 159. For the Spanish text of Mola's threat, see Martínez Bande, *Vizcaya*, p. 23, note 14.
33 Quoted by S. Budiansky, *Air Power* (New York: Viking, 2003), p. 209. See Schüler-Springorum, *La guerra como aventura*, p. 255, note 75. See details of the Condor Legion's Operations room in G. Thomas and M. Morgan-Witts, *The Day Guernica Died* (London: Hodder and Stoughton, 1975), pp. 62–3.
34 Nicholas Rankin, *Telegram from Guernica* (London: Faber and Faber, paperback edition, 2013), p. 157.
35 See H. R. Southworth, *La destruction de Guernica* (Paris, 1975), Chapter 1. See also the most recent Spanish edition (Granada, 2013) with epilogue by Angel Viñas.
36 On Steer and Guernica, see ibid.
37 See *Bombardeo de Cabra* in Wikipedia (accessed 7 March 2018).

38 Thomas and Morgan-Witts, The *Day Guernica Died*, pp. 120–3.
39 Ibid., pp. 197–8.
40 Solé i Sabaté and Villarroya, *España en llamas*, p. 91.
41 X. Irujo, *Gernika 1937: The Market Day Massacre* (Reno: University of Nevada Press, 2015), p. 113.
42 Hooton, *Phoenix Triumphant*, p. 131.
43 Italics added; Von Richthofen's diary quoted by Irujo, *Gernika* 1937, p. 82.
44 Quoted by Solé i Sabaté and Villarroya, *España en llamas*, p. 87. The leader of the raid, Karl von Knauer, reported that the wind changed to coming from the north-east, thus blowing the bombs towards the town (ibid.).
45 Budiansky, *Air Power*, p. 209.
46 Schüler-Springorum, *La guerra como aventura*, p. 256.
47 For discussions at the Non-Intervention Committee about Guernica, see Alpert, *A New International History*, pp. 125–6.
48 Thomas and Morgan-Witts, The *Day Guernica Died*, p. 285.
49 This message is cited by Vicente Talón, *Arde Guernica* (Madrid, 1970), pp. 112–13, probably for the first time in Franco Spain.
50 Schüler-Springorum, *La guerra como aventura*, p. 261.
51 Ibid., p. 263. My thanks to Professor Stephanie Schüler-Springorum for providing the original German phrase.
52 Steer, *The Tree of Gernika*, p. 243.
53 Hugh Thomas, *The Spanish Civil War* (London: Penguin, 2012), quotes the entire letter in his Appendix IV.
54 For this question, see Paul Preston, *The Spanish Holocaust* (London: Harper Press, 2012).
55 Howson, *Aircraft of the Spanish Civil War*, p. 187.
56 Schüler-Springorum, *La guerra como aventura*, p. 262.
57 Ibid., p. 255.
58 Irujo, *Gernika* 1937, p. 47.
59 For an account of the background to the admission of Basque children to the UK, see https://www.basquechildren.org/-/docs/alpert.

Chapter 7

1 This term was used by the insurgents. The Basques called their fortification the *Cinturón defensivo* (information from Nicholas Rankin).
2 James C. Corum, *The Luftwaffe: Creating the Operational Air War, 1918–1940* (Lawrence: University of Kansas Press, 1997), p. 195.
3 Ignacio Hidalgo de Cisneros, *Cambio de Rumbo* (Bucharest, 1964), p. 220.
4 Ibid.
5 Andrés García Lacalle, *Mitos y verdades: la aviación de caza en la guerra española* (Mexico City, 1973), pp. 301–5, 263–4.
6 J. M. Martínez Bande, *Vizcaya* (Madrid, 1971), p. 287.
7 J. Zugazagoitia, *Guerra y vicisitudes de los españoles* (Barcelona, 1977), pp. 266–77.
8 Ferdinando Pedriali, *Guerra di Spagna e aviazione italiana* (Rome, 1992), p. 234.
9 Ibid., p. 235.
10 Ibid., p. 134.

11 See table in ibid., p. 255. Details of these aircraft in Gerald Howson, *Aircraft of the Spanish Civil War, 1936–1939* (Putnam, 1990) (Romeos are listed under 'Imam').
12 Abellán, *Los cazas soviéticos en la guerra aérea de España* (Madrid, 1999), p. 43.
13 See Whelan under names of Soviet airmen.
14 Quoted by Hidalgo de Cisneros, *Cambio de Rumbo*, ii, p. 222.
15 Corum, p. 196.
16 Håkan, 28 September 1937.
17 Ibid.
18 R. Proctor, *Hitler's Luftwaffe in the Spanish Civil War* (New York: Praeger, 1983), p. 165.
19 Abellán, *Los cazas soviéticos en la guerra aérea de España*, Appendix VII.
20 Preston, *Franco* (London: HarperCollins, 1993), pp. 153–4. See also M. Alpert, *La guerra civil española en el mar* (Barcelona, 2007), p. 103.
21 Preston, *Franco*, pp. 82–184.
22 These details have been taken from Howson, *Aircraft of the Spanish Civil War*, pp. 20–8.
23 The highly involved story of this officer's purchase of small civilian aircraft is recounted by Gerald Howson in Chapter 13 of his *Arms for Spain: The Untold Story of the Spanish Civil War* (London: John Murray, 1998).
24 Pedriali, *Guerra di Spagna e aviazione italiana*, p. 246.
25 Howson, *Arms for Spain*, p. 224.
26 All loyal Republican officers were advanced one rank beyond the ranks that they held at the outset of the war.
27 See C. Lázaro Avila, 'Prieto, ministro del Aire', in J. García Fernández, C. Lázaro Avila, A. Puerta Gutiérrez and J. Rodríguez Muñoz (eds), *Indalecio Prieto, primer ministro español del Aire* (Alcalá de Henares, 2016), pp. 59–133, specifically p. 84, note 42.
28 Archives of the Spanish Communist Party 19/7, quoted in *25 militares de la República,* p. 778, note 38.
29 The *Anuario Militar* for 1936 does not usually say where air force officers were stationed. Carlos Engel Masoliver's *El cuerpo de oficiales en la guerra de España* (Valladolid, 2008), unhelpfully, has no index of names, so a particular officer is difficult to trace.
30 Ramón Salas Larrazábal, *Historia del Ejército Popular de la República*, 4 vols. (Madrid, 1973), ii, p. 1503.
31 For Smushkevich's suggestion, see Lázaro Avila in *Indalecio Prieto*, specifically p. 89.
32 Post-war court martials show clearly how few men of Republican convictions were available even in bases such as Cuatro Vientos, Getafe and El Prat (Barcelona), even though the commanders had ensured that the aircraft would remain in the service of the Republic.
33 For these details, see Y. Rybalkin, *Stalin y España: la ayuda militar soviética a la República* (Madrid, 2007), pp. 93–4.
34 Information about the first expedition to Kirovabad led by Major Cascón comes from the essay on him in *25 militares de la República*, specifically pp. 279–82.
35 Andrés Garcia Lacalle, *Mitos y verdades: la aviación de caza en la guerra española* (Mexico City, 1973), pp. 267–9; further information on the bombing of the *Deutschland* comes from the Soviet naval attaché and later Admiral Kuznetsov in *Bajo la Bandera de la España Republicana*, pp. 188–93, and his *Na dalyokom meridiane* (Moscow, 1966), pp. 205–9. For a considered view, see Alpert, *La guerra civil española en el mar*, pp. 296–301.

36 Whelan (under names).
37 On this question, see Indalecio Prieto's allegations in his report to the National Committee of the Socialist Party under the title 'Cómo y por qué salí del ministerio de Defensa' in I. Prieto, *Convulsiones de España*, 3 vols. (Mexico City, 1973), ii, pp. 27–85.
38 García Lacalle, *Mitos y verdades*, pp. 271–9.
39 Pedriali, *Guerra di Spagna e aviazione italiana*, p. 278.
40 See Map IX in J. Salas Larrazábal, La Guerra civil de España. Alpert, in his *The Republican Army in the Spanish Civil War 1936–1939* (Cambridge: Cambridge University Press, 2007), p. 69, shows how the weight of bureaucracy prevented the Republican forces from enjoying the flexibility which characterized the Nationalists. A striking example of this are the twenty-nine documents assembled to deal with the payment of expenses to an air force officer for carrying out a secret mission in Paris and London from May to October 1938 (Archivo General Militar de Avila C.2217.4.62).
41 D. Kowalsky, *La Unión Soviética y la guerra civil española: una revisión crítica* (Barcelona, 2004), p. 299.
42 Quotations in J. M. Martínez Bande, *La ofensiva sobre Segovia y la batalla de Brunete* (Madrid, 1972), pp. 95–6.
43 On this battle, see J. Salas Larrazábal, *La Guerra de España*..., pp. 216–18.
44 On Brunete, see Hugh Thomas, *The Spanish Civil War* (London: Penguin, 2012), pp. 689–90.
45 Håkan's Aviation page referring to July 1937.
46 Confirmed by Whelan (see under Khoziainov and Trusov).
47 Whelan (under surname).
48 From Yakushin's essay in *Bajo la bandera de la España Republicana*, p. 360. I have used Håkan's translation with an emendation (MA).
49 R. K. Smith and R. C. Hall, *Five Down, No Glory. Frank G. Tinker, Mercenary Ace in the Spanish Civil War* (Annapolis, MD: Naval Institute Press, 2011), p. 251.
50 Martínez Bande, *La ofensiva sobre Segovia*, pp. 176–7, notes 202 and 203.
51 Ibid., pp. 219, 227–8.
52 Rybalkin, *Stalin y España*, Table 5, p. 101; and Kowalsky, *La Unión Soviética y la guerra civil española*, pp. 298–9.
53 Smith and Hall, *Five Down, No Glory*, p. 247.
54 Six hundred fifty million lire, according to Pedriali, *Guerra di Spagna e aviazione italiana*, p. 277.
55 Håkan, 18 July 1937.
56 *Ibid*, 21 July, 1937. Lützow evidently did think that the Nationalists were less well supplied with aircraft and pilots than the Republicans.
57 This point is made strongly by Howson in his *Aircraft of the Spanish Civil War*, p. 233.
58 Smith and Hall, *Five Down, No Glory*, p. 258.
59 Ibid., pp. 260–1.

Chapter 8

1 Carlos Engel, *El cuerpo de oficiales en la guerra de España* (Valladolid, 2008), pp. 326–7.

2 For a description of militias in Catalonia and Aragon, see M. Alpert, *The Republican Army in the Spanish Civil War, 1936–1939* (Cambridge: Cambridge University Press, 2013), pp. 49–51.
3 J. Salas Larrazábal, *La Guerra de España*, p. 90. There had been twenty-seven Bréguets at Logroño. Salas does not say how many were made available for the Aragon front.
4 Gerald Howson, *Aircraft of the Spanish Civil War, 1936–1939* (Putnam, 1990), p. 172.
5 J. Salas Larrazábal, *La guerra de España*, p. 91.
6 Ferdinando Pedriali, *Guerra di Spagna e aviazione italiana* (Rome, 1992), p. 264.
7 Ibid., p. 267.
8 Håkan, *Håkan's Aviation Page* (Surfcity.Kund.Dalnet.se/pr32.htm), 26 August 1937.
9 E. Abellán, *Los cazas soviéticos en la guerra aérea de España* (Madrid, 1999), p. 79.
10 Howson, in letter of 25 April 2003 to Frank Schauff and copy to me (MA).
11 Howson, *Aircraft of the Spanish Civil War*, pp. 234–5.
12 R. Proctor, *Hitler's Luftwaffe in the Spanish Civil War* (New York: Praeger, 1983), p. 171.
13 Ibid., p. 172.
14 Pedriali, *Guerra di Spagna e aviazione italiana*, p. 281 and note.
15 Howson, *Aircraft of the Spanish Civil War*, pp. 182–3.
16 Abellán, *Los cazas soviéticos en la guerra aérea de España*, p. 70, note 13.
17 Quotations from Håkan's Aviation Page, January 1938.
18 J. Salas, Larrazábal, *La guerra de España*, p. 286.
19 Stefanie Schüler-Springorum, *La guerra como aventura: la Legión Cóndor en la guerra civil española 1936–1939* (Madrid, 2014), p. 223.
20 Ibid., p. 333.
21 J. Salas Larrazábal, *La guerra de España*, p. 285.
22 Ibid., pp. 211–12.
23 Ramón Salas Larrazábal, *Historia del Ejército Popular de la República* (Madrid, 1973), pp. 1686–7.
24 See Howson, *Aircraft of the Spanish Civil War* under 'Grumman', J. Lario Sánchez, who was selected as a pilot for the Grummans, *Habla un aviador de la* República, (Madrid, 1973), pp. 224–6; and Lacalle, *Mitos y verdades: la aviación de caza en la guerra española* (Mexico City, 1973), pp. 313–15.
25 García Lacalle, *Mitos y verdades*, pp. 283–92.
26 Y. Rybalkin, *Stalin y España: la ayuda militar soviética a la República* (Madrid, 2007) Table 5, p. 101.
27 N. Voronov in *Bajo la bandera de la España Republicana*, pp. 63–131, has an interesting commentary on the Spanish Republican artillery from the point of view of a Soviet adviser.
28 Ramón. Salas Larrazábal, *Historia del Ejército Popular de la República*, p. 1710.
29 Ibid., p. 342.
30 Ibid., pp. 290–3.
31 For discussion of the international situation regarding Spain at this time, see M. Alpert, *A New International History of the Spanish Civil war* (Basingstoke: Palgrave, 2004), Chapter 13.
32 Proctor, *Hitler's Luftwaffe in the Spanish Civil War*, p. 195.
33 Abellán, *Los cazas soviéticos en la guerra aérea de España*, p. 80.
34 Ibid., p. 84.

35 Proctor, *Hitler's Luftwaffe in the Spanish Civil War*, p. 231.
36 García Lacalle, *Mitos y verdades*, pp. 293–5.
37 Pedriali, *Guerra di Spagna e aviazione italiana*, p. 296.
38 J. Salas Larrazabal, *La Guerra de España*, p. 313.
39 Proctor, *Hitler's Luftwaffe in the Spanish Civil War*, pp. 205–6; and Howson, *Aircraft of the Spanish Civil War*, p. 183.
40 *ABC*, Madrid, 29 January 1938.
41 J. Villaroya Font, *Els bombardeigs de Barcelona durant la guerra civil* (Barcelona, 1981), pp. 183–7.
42 Pedriali, *Guerra di Spagna e aviazione italiana*, p. 351.
43 Lacalle, *Mitos y verdades*, p. 308.
44 GD, No. 550, 23 March 1938.
45 J. B. S. Haldane, *A. R. P.* (London: Gollancz, 1938), has many references to the Barcelona bombings.

Chapter 9

1 R. Proctor, *Hitler's Luftwaffe in the Spanish Civil War* (New York: Praeger, 1983), p. 226.
2 J. Larios, *Combat over Spain; Memoirs of a Nationalist Fighter Pilot 1936–1939* (London: Neville Spearman, 1968), p. 217.
3 E. Abellán, *Los cazas soviéticos en la guerra aérea de España*, pp. 115, 120.
4 Ibid., p. 393. Pedriali's list is on his pp. 331–2, Table 8.
5 García Lacalle, *Mitos y verdades: la aviación de caza en la guerra española* (Mexico City, 1973), p. 383. For the camera-equipped Grumman, see p. 385.
6 Abellán, *Los cazas soviéticos en la guerra aérea de España*, p. 131.
7 Ibid., p. 17.
8 Proctor, *Hitler's Luftwaffe in the Spanish Civil War*, p. 224, gives details of Condor Legion missions dated 30 July 1938.
9 Ibid., p. 227.
10 García Lacalle, *Mitos y verdades*, p. 387.
11 Ibid., p. 408.
12 Proctor, *Hitler's Luftwaffe in the Spanish Civil War*, p. 229.
13 Ferdinando Pedriali, *Guerra di Spagna e aviazione italiana* (Rome, 1992), p. 334 and notes 26 and 28.
14 Abellán, *Los cazas soviéticos en la guerra aérea de España*, p. 129.
15 This account is in García Lacalle, *Mitos y verdades: la aviación de caza en la guerra española*, pp. 389–92; and Abellán, *Los cazas soviéticos en la guerra aérea de España*, p. 125.
16 Proctor, *Hitler's Luftwaffe in the Spanish Civil War*, p. 240.
17 Ibid., p. 245.
18 Abellán, *Los cazas soviéticos en la guerra aérea de España*, p. 151
19 Ignacio Hidalgo de Cisneros, *Cambio de Rumbo* (Bucharest, 1964), pp. 242–51.
20 Ibid., pp. 499–500.
21 Juan Lario, *Habla un aviador de la República* (Madrid, 1973), recalls his experiences at the end of the war in pp. 327–40.
22 García Lacalle, *Mitos y verdades*, pp. 496–8.
23 Hugh Thomas, *The Spanish Civil War* (London: Penguin, 2012), p. 869.

24 A full account of the departure of the Republican leaders is given by Paul Preston in *El final de la guerra* (Madrid, 2014), pp. 242–7. However, who flew in which aircraft is unclear. The Dragon Rapide could not have carried more than seven passengers.
25 The episode is very thoroughly discussed also by Luis Romero, *El final de la guerra* (Barcelona,1976), pp. 321–3.
26 Abellán, *Los cazas soviéticos en la guerra aérea de España*, Appendix X.
27 S. Casado, *The Last Days of Madrid* (London: Peter Davies, 1939), p. 254.
28 Pedriali, *Guerra di Spagna e aviazione italiana*, p. 368. According to G. Howson, *Aircraft of the Spanish Civil War* (Putnam, 1990), pp. 234–5, only one Bf.109 E-1 arrived before mid-December 1938. This aircraft had a top speed of 285 mph, 68 mph more than the previous version.
29 Ballester told his story to Andrés García Lacalle, presumably in Mexico, and the latter repeats it in his book (pp. 448–69).
30 www. portal/cultura.mde.es/Galerias/cultura/archivo/fichero/096 (accessed 6 December 2016).
31 Montilla's obituary by Juan M. Riesgo appeared in the daily newspaper, *El País* (Madrid, 2 June 2007).
32 C. Lázaro Avila, *25 Militares de la República* (Madrid, 2011), pp. 540–2.
33 See Whelan's alphabetical list.
34 Pedriali, *Guerra di Spagna e aviazione italiana*, p. 378, note 38.
35 Stefanie Schüler-Springorum, *La guerra como aventura: la Legión Cóndor en la guerra civil española 1936–1939* (Madrid, 2014), pp. 325–34.
36 Ibid., p. 334
37 Ibid., p. 361.
38 Ibid., pp. 376–7.

Chapter 10

1 Stefanie Schüler-Springorum, *La guerra como aventura: la Legión Cóndor en la guerra civil española 1936–1939* (Madrid, 2014), p. 116.
2 Ibid., p. 111.
3 Ibid., p. 145.
4 J. Larios, *Combat over Spain: Memoirs of a Nationalist Fighter Pilot 1936–1939* (London: Neville Spearman 1968), p. 88ff.
5 E. Hooton, *Phoenix Triumphant: The Rise and Fall of the Luftwaffe* (London: Brockhampton Press, 1999), p. 125.
6 Ibid., p. 155.
7 Schüler-Springorum, *La guerra como aventura*, p. 170. My thanks to Professor Schüler-Springorum for providing the original German quotation.
8 Ibid., p. 183.
9 Ibid.
10 Ibid., p. 185.
11 Ibid., p. 195.
12 Ibid., p. 198.
13 P. Preston, *Franco* (London: HarperCollins, 1993), p. 473.
14 Virginia Cowles, *Looking for Trouble* (London: Hamish Hamilton, 1941), pp. 68, 86.
15 My thanks to Giancarlo Garello and Angelo Emiliani for this information.

16 R. K. Smith and R. C. Hall, *Five Down, No Glory. Frank G. Tinker, Mercenary Ace in the Spanish Civil War* (Annapolis, MD: Naval Institute Press, 2011), p. 259.
17 See Commissar Krotov's report in R. Radosh, M. R. Habeck and G. Sevostianov (eds), *Spain Betrayed: The Soviet Union and the Spanish Civil War* (New Haven, CT: Yale University Press), Document 59. See also D. Kowalsky, *La Unión Soviética y la guerra civil española: una revisión crítica* (Barcelona, 2004), p. 328.
18 Cowles, *Looking for Trouble*, pp. 83–4.
19 See R. Hallion, *Strike from the Air: The History of Battlefield Air Attack 1911–45* (Shrewsbury: Airlife, 1989), p. 113.
20 I have used the description of the 'Finger-Four' formation given by Len Deighton in his *Fighter: The True Story of the Battle of Britain* (London: Pimlico, 1993), pp. 130–1.
21 K. Ries and H. Ring, *The Legion Condor* (West Chester, PA: Schiffer, 1992). Translated from Karl Ries and Hans Ring, *Legion Condor, 1936–1939. Eine illustrierte Dokumentation* (Mainz,1980), pp. 234–5.
22 See M. Haapamaki, *The Coming of the Aerial War: Culture and the Fear of Airborne Attack in Inter-War Britain* (London: I.B. Tauris, 2014), p. 92.
23 Y. Rybalkin, *Stalin y España: la ayuda militar soviética a la República* (Madrid, 2007), pp. 148–9.
24 Ibid., pp. 158–61.
25 Kowalsky, *La Unión Soviética y la guerra civil española*, p. 297.
26 Ibid., pp. 222, 290, 295–6.
27 Rybalkin, *Stalin y España*, Table 5.
28 Indalecio Prieto 'Cómo y por qué salí del ministerio de Defensa Nacional: intrigas de los rusos en España', especially p. 65. See also the letter sent by Lieutenant Colonel Angel Riaño, Herad of the Information Section of the Air Staff, to Colonel Camacho, complaining about the lack of communication with the Russian commanders, in J. Salas Larrazábal, *Guerra Aérea 36/9* (Madrid, 1998), Vol. 1, Appendix 65.
29 See Whelan under 'Lopatin'.
30 Ibid., pp. 166–7.
31 Edoardo Grassia, '"Aviazione Legionaria": il comando strategico politico e tecnico-militare delle forzee aeree italiane impegnate nel conflitto civile spagnolo', *Diacronie: Studi di Storia Contemporanea* 7, 3 (2011), pp. 1–23.
32 E. Mastrorilli, 'Guerra civile spagnola, intervento italiano e guerra totale' *Revista Universitaria de Historia Militar* 3, 6 (2014), pp. 68–86, particularly pp. 77–81. I thank the author for sending me this article.
33 Stephen S. Budiansky, *Air Power* (New York: Viking), 2003, p. 212. Much useful information which can put the Spanish Civil War's air war into focus can be found in Patrick Bishop's *Battle of Britain* (London: Quercus, 2010).

Bibliography

Primary sources

1. Documents on Spanish Republican and Nationalist air forces held in the Archivo General Militar, Avila.
2. *Documents Diplomatiques Français*, Series 2 (1936–9), Vol. 3 No. 46 (Paris, Imprimerie Nationale, 1966).
3. *Documents on British Foreign Policy 1919–1939*, 2nd Series, Vol. 17 (HMSO, London, 1979).
4. *Documents on German Foreign Policy 1918–1945*, Series D, Vol. 111 (HMSO, London, 1951).

Books and articles (the editions listed are the ones which have been used)

Abella, R., *La vida cotidiana durante la guerra civil: la España nacional* (Barcelona,1973).
Abellán, E., *Los cazas soviéticos en la guerra aérea de España* (Madrid, 1999).
Air Ministry, *The Rise and Fall of the German Air Force 1933–45* (London: Air Ministry, 1948).
Alpert, M., *A New International History of the Spanish Civil War*, 2nd ed. (Basingstoke: Palgrave, 2004).
Alpert, M., *La guerra civil española en el mar* (Barcelona, 2007).
Alpert, M., *La reforma militar de Azaña 1931–1933*, 2nd ed. (Granada, 2008).
Alpert, M., 'The Clash of Spanish Armies: Contrasting Ways of War in Spain 1936–1939', *War in History* 6, 3 (1999), pp. 331–51.
Alpert, M., *The Republican Army in the Spanish Civil War, 1936–1939* (Cambridge: Cambridge University Press, 2013).
Andersson, L., *Soviet Aircraft and Aviation* (London: Putnam, 1994).
Arrarás, J., *Historia de la Cruzada Española*, Vol. 3, Book XI (Madrid, 1941).
Bajo la bandera de la España republicana: recuerdan los voluntarios soviéticos participantes en la guerra nacional-revolucionaria en España (Moscow, no date [1967?]).
Baker, D., *Adolf Galland: The Authorised Biography* (London: Windrow and Greene, 1996).
Balfour, S., *Deadly Embrace: Morocco and the Road to the Spanish Civil War* (Oxford: Oxford University Press, 2002).

Belmonte Díaz, J., *Avila en la guerra civil* (Bilbao, 2013).
Boyd, A., *The Soviet Air Force since 1918* (London: MacDonald and Jane's, 1977).
Brenan, G., *Personal Record 1920–1972* (New York: Knopf, 1975).
Bridgeman, B., *The Flyers* (Swindon: Brian Bridgeman, 1989).
Budiansky, S., *Air Power* (New York: Viking, 2003).
Cabrera, M., *Juan March 1880–1962* (Madrid, 2011).
Cerdera, L. M., *Málaga, base naval accidental* (Seville, 2015).
Ceva, L., 'Conseguenze dell'intervento italo-fascista', in Sacerdoti Mariani, G., Columbo, A. and Pasinato, A. (eds), *La guerra civile spagnola tra politica e letteratura* (Florence, 1995), pp. 215–29.
Chalmers-Mitchell, P., *My House in Málaga* (London: Faber and Faber, 1938).
Ciano, G., *Ciano's Hidden Diary 1937–1938* (New York: Dutton, 1953).
Cooper, B., *The Story of the Bomber 1914–1945* (London: Octopus, 1974).
Cooper, M., *The German Air Force 1933–1945* (London: Jane's, 1981).
Corum, J. S., 'The Luftwaffe and the Coalition Air War in Spain 1936–1939', *Journal of Strategic Studies* 18, 1 (1995), pp. 68–90.
Corum, J. S., *The Luftwaffe: Creating the Operational Air War 1918–1940* (Lawrence: University Press of Kansas, 1997).
Corum, J. S., 'The Spanish Civil War: Lessons Learned and Not Learned by the Great Powers', *Journal of Military History* 4, 62 (1998), pp. 313–34.
Coverdale, J. F., *Italian Intervention in the Spanish Civil War* (Princeton, NJ: Princeton University Press, 1975).
Dary, Jean, 'La Guerre Aérienne en Espagne', *Revue de l'Armée de l'Air* 96 (1938), pp. 808–24.
Davies R. W., et al. (eds.), *The Stalin–Kaganovich Correspondence, 1931–1936* (New Haven, CT: Yale University Press, 2003).
Deighton, L. *Fighter: The True Story of the Battle of Britain* (London: Pimlico, 1993).
De Wet, O., *Cardboard Crucifix* (London: Blackwood, 1938).
Edwards, J., *Airmen without Portfolio: US Mercenaries in Civil War Spain* (Westport, CT: Praeger, 1997).
Engel Masoliver, C., *El cuerpo de oficiales en la guerra de España* (Valladolid, 2008).
Friends of Spain, *The Spanish War: Foreign Wings over the Basque Country* (London: The Friends of Spain, 1937).
Fustas Martí, S., *Vivencias de un piloto-aviador 1936–1939* (Terrassa, 2006).
Galland, A., *The First and the Last: The German fighter force in World War II* (London: Methuen, 1955).
García Fernández, J. (ed.), *Veinticinco militares de la República* (Madrid, 2011).
García Fernández J., Lázaro Avila, C., Puerta Gutiérrez, A. and Rodríguez Muñoz, J., *Indalecio Prieto, primer ministro español del Aire* (Madrid, 2016).
García Lacalle, Andrés, *Mitos y verdades: la aviación de caza en la guerra española* (Mexico City: Oasis, 1973).
García Morato, Joaquín, *Guerra en el aire* (Madrid, 1940).
Gisclon, Jean, *Des avions et des hommes* (Paris, 1969).
Gooch, J., *Mussolini and His Generals: The Armed Forces and Fascist Foreign Policy, 1922–1940* (New York: Cambridge University Press, 2007).
Grassia, Edoardo, '"Aviazione Legionaria": il comando strategico politico e tecnico-militare delle forzee aeree italiane impegnate nel conflitto civile spagnolo', *Diacronie: Studi di Storia Contemporanea* 7, 3 (2011), pp. 1–23.
Green, William, *Augsburg Eagle: The Story of the Messerschmitt 109* (London: MacDonald, 1971).

Griner, M., *I Ragazzi del '36* (Milan, 2006).
Haapamaki, M. *The Coming of the Aerial War: Culture and the Fear of Airborne Attack in Inter-War Britain* (London: I.B. Tauris, 2014).
Håkan's Aviation Page (the Spanish Civil War by year) at <surfcity.kund.dalnet.se/scw-*year*.htm> (last consulted 5 February 2019).
Haldane, J. B. S., *A. R. P.* (London: Gollancz, 1938).
Hall, R. C. and Smith, R. K., *Five Down, No Glory. Frank G. Tinker, Mercenary Ace in the Spanish Civil War* (Annapolis, MD: Naval Institute Press, 2011).
Hallion, R., *Strike from the Sky: The History of Battlefield Air Attack 1911–1945* (Shrewsbury: Airlife, 1989).
Hidalgo de Cisneros, Ignacio, *Cambio de Rumbo*, Vol. 2 (Bucharest, 1964).
'Hispanicus', *Foreign Intervention in Spain*, Vol. 1 (London: United Editorial, no date).
Hooton, E., *Phoenix Triumphant: The Rise and Fall of the Luftwaffe* (London: Brockhampton Press, 1999).
Howson, G., *Aircraft of the Spanish Civil War, 1936–1939* (London: Putnam, 1990).
Howson, G., *Arms for Spain: The Untold Story of the Spanish Civil War* (London: John Murray, 1998).
Irujo, X., *Gernika 1937: The Market Day Massacre* (Reno, NV: University of Nevada Press, 2015).
Jackson, R, *The Red Falcons: The Soviet Air Force in Action 1919–1969* (London: Clifton Books, 1970).
Keene, J., *Fighting for Franco* (London: Leicester University Press, 2001).
Kilmarx, R., *A History of Soviet Air Power* (London: Faber and Faber, 1962).
Koestler, A., *Dialogue with Death* (London: Collins and Hamish Hamilton, 1954).
Koestler, A., *Spanish Testament* (London: Gollancz, 1937).
Koltsov, M., *Diario de la guerra de España* (Paris, 1963).
Kowalsky, D., *La Unión Soviética y la guerra civil española: una revisión crítica* (Barcelona, 2004).
Kuznetsov, N., *Na dalyokom meridiane* (Moscow, 1966).
Langdon-Davies, J., *Air Raid* (London: Routledge, 1938).
Lario Sánchez, J., *Habla un aviador de la República* (Madrid, 1973).
Larios, J., *Combat over Spain: Memoirs of a Nationalist Fighter Pilot 1936–1939* (London: Neville Spearman, 1968).
Lázaro Avila, C., 'Ignacio Hidalgo de Cisneros', in *25 militares de la República* (Madrid, 2011), pp. 503–43.
Martínez Bande, J. M., *La campaña de Andalucía* (Madrid: Servicio Histórico Militar, 1969).
Martínez Bande, J. M., *La invasión de Aragón y el desembarco en Mallorca* (Madrid, 1970).
Martínez Bande, J. M. *La ofensiva sobre Segovia y la Batalla de Brunete* (Madrid, 1972).
Martínez Bande, J. M., *Vizcaya* (Madrid, 1971).
Mason, H. M., *The Rise of the Luftwaffe* (London: Cassell, 1975).
Mastrorilli, E, 'Guerra civile spagnola, intervento italiano e guerra totale', *Revista Universitaria de Historia Militar* 3, 6 (2014), pp. 68–86.
Matteoli, G., *L'Aviazione Legionaria in Spagna* (Rome, 1940).
Muggeridge, M. (ed.), *Ciano's Diary 1939–1943* (London: Heinemann, 1947).
Murray, W, *The Luftwaffe: Strategy for Defeat 1933–45* (Washington, DC, 1996).
Northomb, P, *Malraux en Espagne* (Paris, 1999).
Patterson, I., *Guernica and Total War* (Cambridge, MA: Harvard University Press, 2007).
Pedriali, F., *Guerra di Spagna e aviazione italiana* (Rome, 1992).

Pelliccia, A., *La Regia Aeronautica dalle origini alla seconda guerra mondiale 1923–1943* (Rome, 1992).
Pérez de Sevilla y Ayala, F., *Italianos en España* (Madrid, 1958).
Pérez San Emeterio, C., *Historia de la Aviación española* (Madrid, 1988).
Preston, P., *Franco* (London: HarperCollins, 1993).
Preston, P., *The Spanish Holocaust* (London: Harper Press, 2013).
Proctor, R., *Hitler's Luftwaffe in the Spanish Civil War* (New York: Praeger, 1983).
Radosh, R., Habeck, Mary R. and Sevostianov, G. (eds), *Spain Betrayed: The Soviet Union and the Spanish Civil War* (New Haven, CT: Yale University Press, 2001).
Rankin, N., *Telegram from Guernica* (London: Faber and Faber, 2013).
Ries, K. and Ring, H., *The Legion Condor* (Transl. D. Johnston) (West Chester, PA: Schiffer, 1992.
Romero, Luis, *El final de la guerra* (Barcelona, 1976).
Romero, Luis, *Tres días de julio* (Esplugues de Llobregat, 1967).
Rybalkin, Y., *Stalin y España: la ayuda militar soviética a la República* (Madrid, 2007).
Salas Larrazábal, J., *Guerra Aérea 36/9*, 4 volumes (Madrid, 1998).
Salas Larrazábal, J., *La guerra de España desde el aire* (Esplugues de Llobregat, 1969), translated by Margaret Kelley as *Air War over Spain* (London: Ian Allan, 1974).
Salas Larrazábal, R., *Historia del Ejército Popular de la República*, 4 vols. (Madrid, 1973).
Sánchez Ruano, F., *Islam y la guerra civil española* (Madrid, 2004).
Schüler-Springorum, Stefanie, *La guerra como aventura: la Legión Cóndor en la guerra civil española 1936–1939* (Madrid, 2014), translated from *Krieg und Fliegen: die Legion Condor im Spanischen Bürgerkrieg* (Paderborn, 2010).
Skoutelsky, R., *Novedad en el frente* (Madrid, 2006).
Smith, M. A., *British Air Strategy between the Wars* (Oxford: Clarendon Press, 1984).
Solé i Sabaté, J. and Villaroya, J., *España en llamas: la guerra civil desde el aire* (Madrid, 2003).
Solé i Sabaté, J. and Villaroya, J., *Catalunya sota les bombes (1936–1939)* (Abadía de Montserrat, 1986).
Southworth, H., *Guernica!,Guernica! A study of Journalism, Diplomacy, Propaganda and History* (Berkeley: University of California Press, 1977). Spanish edition with postscript by Angel Viñas (Granada, 2013).
Steer, G. L., *The Tree of Gernika: A Field Study of Modern war* (London: Hodder and Stoughton, 1938).
Sullivan, B., 'Downfall of the Regia Aeronautica', in Higham, R. and Harris, Stephen J. (eds), *Why Air Forces Fail: The Anatomy of Defeat* (Lexington: University Press of Kentucky, 2006), pp. 135–78.
Sullivan, B., 'Fascist Italy's Military Involvement in the Spanish Civil War', *Journal of Military History* 59, 4 (1995), pp. 697–727, especially p. 718.
Sullivan, B., 'The Italian Armed Forces 1918–1940', in Millet, A. R. and Murray, W. (eds), *Military Effectiveness*, Vol. 2, 3 vols. (Boston, MA: Allen and Unwin, 1987), pp. 169–217.
Talón, V., *Arde Guernica* (Madrid, 1970).
Thomas, G. and Morgan-Witts, M., *The Day Guernica Died* (London: Hodder and Stoughton, 1975).
Thomas, H., *The Spanish Civil War* (London: Penguin, 2012).
Thornberry, R. S., *André Malraux et l'Espagne* (Geneva, 1977).
Tinker, F., *Some Still Live: Experiences of a Fighter-Plane Pilot in the Spanish War* (London: Lovat Dickson, 1938).
Trevor-Roper, H. R. (ed.), *Hitler's Table Talk* (London: Weidenfeld and Nicolson, 1953).

Vaquero Peláez, D., *Creer, obedecer, combatir…y morir. Fascistas italianos en la guerra civil española* (Zaragoza, 2006).
Vila San Juan, J. L, *Enigmas de la guerra civil española* (Barcelona, 1974).
Villaroya Font, J., *Els bombardeigs de Barcelona durant la guerra civil* (Abadía de Montserrat, 1981).
Viñas, A., *El oro de Moscú* (Barcelona, 1979).
Viñas, A., *Hítler, Franco y el estallido de la guerra civil; antecedentes y consecuencias* (Madrid, 2001).
Viñas, A., 'La connivencia fascista con la sublevación y otros éxitos de la trama civil', in Viñas, A. et al. (eds), *Los mitos del 18 de julio* (Barcelona, 2013), pp. 79–181.
Viñas, A. and Collado Seidel, C. 'Franco's Request to the Third Reich for Military Assistance', *Contemporary European History* 11, 2 (2002), pp. 191–210.
Wake, Jehane, *Kleinwort Benson: The History of Two Families in Banking* (Oxford: Oxford University Press, 1997).
Wheatley, R., *Hitler and Spain: The Nazi Role in the Spanish Civil War* (Lexington: University Press of Kentucky, 1989).
Whelan, P., *Soviet Airmen in the Spanish Civil War* (Atglen, PA: Schiffer, 2014).
Yakovlev, A. S., *The Aim of a Lifetime* (Moscow: Progress, 1972).
Zugazagoitia, J., *Guerra y vicisitudes de los españoles* (Barcelona, 1977).

Index